SKELMERSDALE

FICTION RESERVE STOCK £L.60

AUTHOR	CLASS
MCEWAN, T.	F

TITLE

Arithmetic.

ARITHMETIC

Todd McEwen's previous novels are *Fisher's Hornpipe* and *McX*.

Todd McEwen

ARITHMETIC

V

VINTAGE

07727459

Published by Vintage 1999

2 4 6 8 10 9 7 5 3 1

Copyright © Todd McEwen 1998

First published in Great Britain by
Jonathan Cape in 1998

Vintage
Random House, 20 Vauxhall Bridge Road,
London SW1V 2SA

Random House Australia (Pty) Limited
20 Alfred Street, Milsons Point, Sydney
New South Wales 2061, Australia

Random House New Zealand Limited
18 Poland Road, Glenfield,
Auckland 10, New Zealand

Random House South Africa (Pty) Limited
Endulini, 5A Jubilee Road, Parktown 2193, South Africa

Random House UK Limited Reg. No. 954009

A CIP catalogue record for this book
is available from the British Library

ISBN 0 09 977261 2

Printed and bound in Great Britain by
Cox & Wyman, Reading, Berkshire

I live in an invented place whose only purpose is avoidance, and what I would avoid, I carry with me, always.

JOHN BURNSIDE, *Suburbs*

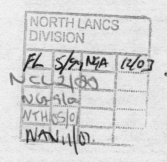

TO KIRSTY

ARITHMETIC

M om and Dad lived in equanimity, in a happy marriage, until one day when these two guys knocked on the front door. They had the desert intellectual look, geology sunglasses, white shirts, no ties. One wore a cap and the other a hat. The one in the cap represented the County of Orange, and the one in the hat Walt Disney Productions. May we come in?

My mother's tears began to trickle, then ran freely, as they put to her the idea of Disneyland, what a wonderful thing, and although they acknowledged it to be unfortunate for the few, in fact an extremely bum deal, grant you that, the County of Orange thought it best to exercise its right of eminent domain in this case. Disneyland would ensure the perpetuation and florescence of the County of Orange for eternity, as well as provide for the tuition and entertainment of a whole new generation of American children and their families. No choice anyway lady, sign here.

When Dad got home that night, still optimistic, still in a scientific freshness from his job of a year, he found Mom lying on the sofa with the canceled quit-claim to their little piece of California in her hand. There was a crumpled promissory letter from Walt Disney's big fat brother and the mayor of the next town, offering to buy them a house there. These pages Mom had fired briefly, in the afternoon, but had thought better of it. They were still legible though they had tears and maraschino cherry juice on them.

I'm home, said Dad.

Is this America or not, Mom wanted to know. She had invested a lot of time, and money, in the back yard, in flowers, in sewing an awning out of dark green canvas. She'd spent so much time in simply thinking of this as her home.

In the end of course they had to move, rarely has there been a government more brutish and menacing than the County of Orange, and their new town was, in some ways, fresher. But in her new tract home, for which she had given up her Spanish bungalow with the red concrete walk, Mom began to think about Walt Disney and to plot against him.

Look at these books, she said. Why he changes the stories all around. That crook, that king! He exploits our fears, have you ever seen anything more coldly calculating than the fire scene in *Dumbo*. And he coats everything with—

Her greatest fear was that he would move on *Little Women*, the literary treasure of her child-hood, her *girlhood*, as she always called it, animate it with lounge music and the voices of entertainers

4

she thought were creepy. Women who led cigarette and lipstick lives, and men who were all trouser. She had nightmares about it. Then would come the Little Women RIDE, what would it be like, neat cottages with SUITORS in Currier & Ives pants, God, it didn't bear thinking about; she couldn't help it.

No harm in going, Dad said, when Disneyland opened. *No, no, no*, she said, but he made her go. She was very judgmental and haughty at the gates, Dad said, but seemed to calm and to cool on the jungle boat ride. Bongo music, the white boat with a striped canvas roof, low in the water. They chugged under a waterfall, the plash of which mingled with the pleasant fussing of midwest visitors in spray, *Oh, how, well, authentic, Mary Lou*, and rounded a bend. A hippopotamus head came out of the water and everyone said *Ooo!*

Don't worry, folks, said the college boy jungle guide in his white pants with a big revolver. He aimed at the marauding head. They said *Ooo!* and stuck their fingers in their ears, when all of a sudden Mom jumped up and pointed down into the water (made murky to hide hippopotamus machinery) yelling, *There! Right there, that's where my house was, you miserable Philistines*. Everyone turned to look at her as Dad jumped up and held her in his arms.

I had the most beautiful home there, look, you can just see part of the walk, she said. The midwest ladies were interested in this plaint, the men

5

more discomfited. *Right by that window was the breakfast nook*, Mom went on, *the very nook where they came, from Walt Goddamn Disney, and made me sign away my pretty little house for this, this . . . chicken-wire Congo*.

The jungle guide stared at them, without thinking, pointing his gun at Mom. Ohio went for cover under the cushions. *Are you all right, ma'am*, said the jungle guide. *Put the gun down, sonny*, said Dad, *nice and slow*. He had always wanted to say something like that. The jungle guide looked down at the gun in his hand and yelled in surprise. He flung the gun out into the river where it bounced off the head, which had been waiting, mouth open, to be shot. *Boat five*, said the intercom, *you have not moved past hippo switch*.

When Mom got off the boat, still dabbing at her eyes, she rounded on the poor college boy and choked, inexplicably, *this is just like pumping that goo through Lenin, he never deserved that*.

6

His nose, however, again gushed out blood, a system of defence which seemed as natural to him as that resorted to by the race of stinkards.

JAMES HOGG, *The Private Memoirs and Confessions of a Justified Sinner*

I was glad, I said once, finding I was tall enough to place my chin on the cool tile of the kitchen counter, that I didn't have to go to school. Her shock. *Oh, but school is*—I know. I'm just glad I don't have to go yet.

I didn't usually remember that. I woke up and looked at the dumb cowboys on my walls and my curtains. Down in the non-Science end of the neighborhood Fard would be getting up out of his turmoily bed, I thought. He'd stand there rubbing his eyes in his room, which wasn't entirely his room, looking at his big brother's junk, where was he, white socks with Lambretta grease, black slip-ons. Fard would be thinking about grooming himself with teenage products in the bathroom, Fard, standing there in the roar of the machines, awakened and now rooted in his own existence by laundry. His mom took it in. But did she ever give it back, is what we wanted to know.

Fard's mom's face was like laundry, folded but not ironed. All their faces were folded, by having to share rooms, by having a folded grandmother in the house, a blurry tv, folded by laundry and

7

the idea of laundry. When you went to Fard's, laundry was what was going on. They measured their lives by loads of it, meals were sandwiches eaten between it and on top of it. They couldn't talk without yelling over the washer and dryer, Fard's mom was hoarse.

Laundry drove Fard's family from their home. Fard never wanted to be there, and his brother lived on the driveway, a scooter tinker. Fard's backyard was a desert, guarded by their dog, only ever seen as a frantic snout. They parked a trailer in their front yard, but never took a vacation in it, it was more like their family room. The trailer, I thought, its baked, resinous smell. Fard's little sister, her swimsuit, one summer day—she showed us.

Now Fard will have found some of his own clothes among the piles and piles of laundry. He'll wash his face with Noxzema (teenage products are the only ones in his bathroom, Vitalis, rubber rollers) which added to his margariney aspect, comb his Regular Boy's exactly, watching his own face worry already in the mirror, call to his mom *goodbye* over the machines and walk up the street to my house.

When you just get along with a guy.

Big fin rockets, giant wheel space stations made their own gravity on Fard's lunch box. I had the same box, we all saw it at the store during the summer. Nunzio even had the thermos, with SATURN on it, which our moms wouldn't buy, *you'll only break it*. Nunzio also had official pants for everything, thousands of ping-pong balls, a million rolls of caps, fabulous wealth beyond your wildest dreams, why with a friend like that you could RULE THE WORLD. Nunzio stank.

Fard said hello in his polite way to my Mom and we looked at each other in our jeans, still new. Fard's jeans were bought very long and rolled most of the way to his knees to save money. They hadn't even been washed yet, hadn't become laundry, he walked like the Tin Man. Fard looked a little hot, too, every 'autumn' they give him a foody-looking cardigan.

Are you boys all ready? You look very nice, Fard.

Thank you, Mrs Lake.

We walked in the heat like old men, stumping in our stiff clothes. Usually we started talking about space right away, but Fard was downcast.

Things aren't so hot, he said before Mom even stopped waving to us, down at the Broaster.

Fard's dad bought a Broaster. Laundry drove him out of their house and in his slow truck wanderings he found another something time-consuming, hot, humid, noisy. A cheap hot-fat animal funeral, just the thing. So then despite the washing machine and dryer thundering day and night at home, Fard's mom and sisters spent a lot of time at their Broaster, one of a little row of shops way out on Orangethorpe. Fard's dad worked there sometimes, but kept circling the county in his truck.

His smile and words came slow. His eyes moved slowly, his truck was slow and its gate and doors were slow. Fard rode his bike slowly, and it was the same rust-red and white of the slouching truck.

Fard's dad slowly took us to a carnival once. It was marooned at a tiny shopping center out in the orange groves. It was nine-thirty on a Saturday morning, Fard's dad would have to work later, so I thought, in a huge, farm-like mine where he crawled with friendly earthworms and carrots and that was how he got dirt into all the creases of his clothes and the folds of his face. The carnival was barely open that early. Fard's dad contented himself walking around just looking at the rides. I could see Fard starting to stew about the slow family truck, his father's folded look, and visiting a carnival when you could still see how they put things together—finally and boldly he asked for fifty cents for the Ferris wheel.

Fard's dad reached slowly for his back pocket, he wore jeans like we did, rolled up like a kid's, like Fard's, except the cuffs were filled with loam, and handed the money to Fard like he was giving him LIFE. There was organism, and love, in that one slow handing of a dollar, one with the slow-forming wrinkle under Fard's dad's eye and the dawning of one of his teeth—like watching a hill smile. The money came from the dirt and the carrots and the

10

slow truck and the suffering Broaster. Fard was hopelessly caught up in it, so many conversations on the way home in the truck, over the late-night folding of laundry. But Fard laughed as he took the dollar from his dad. Fard's laugh was high and often cracked with worry; he worried that laughing might cost his family something.

After the carnival we went to the Broaster. I felt important standing in a place with hundreds of chickens, thousands of potatoes, cases of pop, my first look behind the scenes of retail. Fard's mom and his sisters bent over, doing things in the fatty heat, the way they looked at home among laundry.

Hello, boys, listlessly.

Can I show Joe around? Can we have potatoes?

A long folded look from his mom. I'll put a couple in for you. Do you want to share a 7-Up?

Wow, do we.

Black knobs on the sealing levers of the broasting machines interested me, but Fard had lost any feeling of intrigue about the place, worried as he was about its survival. I joked with him about broasting various things, maybe Nunzio, but this caused him only pain. His mom and sisters were looking at us funny and I realized Fard was there to work, that he would stay late into the night broasting things and putting them into foil bags. Later their whole life would smell like this, not even laundry could stop it.

I tried to interest Mom and Dad in the new science of broasting, but they wouldn't have it.

I'm not driving all the way out Orangethorpe. You're not eating that.

Fard and I crossed the Big Street. He still looked worried. There was a row of apartment houses, *Bali Hai*, *Flamingo*, *Vista View*, *O'Connor*, on the left-hand side of the street. Bright stucco fronts, some of them flagged, palm trees, palm trees. From most of these places came the patio shimmer that meant *pool*.

There are pools and there are pools, I thought, some bluer and warmer than the most fabulous motel pool of them all. Some, where people live, are like ashtrays, they have TURDS floating in them.

I thought of the road to school as TREE-LINED and straight, as having a vanishing point you could clearly see, not lost under the haze and the glare of cars. Thanks to the 'weather' the view was always the same and as I daily crossed the Big Street and came upon this shot, I rolled the opening credits to my show, about Anita and me. Chummy music, winsome oboes and a little brass.

I'm still not sure what the credits should say, I said to Fard, but they're big.

Big juicy yellow titles, said Fard, like Disney.

Perhaps just *Anita*. In the title shot she walked toward the camera and stood, smiling. The credits, the music, the shot toward school, the opening of my show was the best part of my day. Although Anita liked me she sometimes ignored me, then I became aware of the urgent pressures of our class. Pressures of arithmetic, of Gomez, putting his finger—

I can't keep track, I said, of who I like and who shouldn't be liked, anymore. Or who likes me.

If that's true, said Fard, they'll dump all over you.

The crossing guard, his cap of yellow leather, wrinkled exactly like his face. Dragged himself out of his Bali Hai blue Bel Air and held up his freshly painted paddle.

On the first day of school we stopped right here in the middle of the street as Fard had said, *Oh, man, I forgot to tell you—we have Her!* There we were in front of school for another year, and I wanted to grab Fard. While I had been on vacation, playing with magnetic games, looking at thunderstorms, fighting with Julie, throwing up, FARD was writhing nights in the salad of his bedclothes, nearly torn apart by the knowledge we had been given Favorite Teacher, he was practically ruined by it by the time we got back into town. The knowledge that we had nearly killed Fard, by going on vacation, swept over me. He could barely croak it out that morning, he needed medical help.

All last year we rescued Her. Favorite Teacher. We talked about Favorite Teacher, before we knew Her name, Mrs Dentyne, late into the evenings. I would sit on the swing while Fard pumped the glider a little. Or we were in Fard's room, our thoughts in turmoil from its turmoil, out of the textile jungle would come thoughts of Her, our next mission, all for Her. Just thinking about Her made us both quiet. I got sofa feelings.

We had a sofa at home, soft and billowed as the word sofa. When you *crawled behind the sofa*, that was what gave you the best ball-tickle and raise-hackle. Here was an instant place, a place always waiting, always gloomy. To know I could crawl behind the sofa gave me a secret mighty feeling.

I started looking at all our furniture as places, living life on a lower level. On the black camphorwood chest was the scene of a river, boats floating in a swamp, thick TREES right to the edge of the water. I was afraid I would get drawn in to this, I'd stare at it until sofa feeling began and then go off on my hands and knees.

Behind the sofa became my usual spot, it was the first place Mom looked for me and when Dad came home he'd start talking to the sofa. He spoke to it perfectly naturally, addressing me once when we had company and I wasn't behind it.

Let's say this guy from another planet CAPTURES Favorite Teacher, while She's doing Scientific work, and takes her back there—

And we only have twenty-four hours to rescue Her, said Fard—

Before he gives Her a shot—

Which makes Her his!

14

Yeah!

It got so we didn't have to get up and act it out, we had this new mighty feeling which put an end to that, we rarely had to leave the stage we set. And now we really did have Her, I thought on the first day.

Would you get out of the crosswalk, said the guard.

Fard and I had been staring at each other with our mouths open. And then school's mouth opened, the cool arcade where the bathrooms waited and the drinking fountains dripped. Back, we were back, with Julius and Gomez and Kurt, and Anita galloping around the Big Tree. New sixth graders in red sashes told you to walk, *kid*.

The bell, sounding of autumn in summer. We went to line up in front of our new rooms. Some happy, delighting in totems of the new year, canvas binders, lunch boxes, stiff jeans, the Principal's bow tie throbbing above his white shirt, being pushed around by his larynx. Some were realizing, only now, in lining up, that summer was over, the chill of chalk was in their blood, goodbye to running in and out of sprinklers, leisurely living the whole days of their blocks, milkman, egg lady, bakery van, mailman, streetsweeper, ice-cream man, air conditioners and the chug of watering going on when you woke up, and the talk of adults and their barbecues when you went to lie beneath your sheet, their shadows on the window shade—and these were already kicking hell out of each other.

Some couldn't wait to cry out at the first injustice of the classroom, to think up horrors to tell each other in the toilets.

Seething with hope, love and the alphabet, Science, milk, outrage and venom: the long, low, olive-green school.

So, said Gomez, descending on us, there's a space launch today. You got your little mama's-boy radio?

Space *mission*, said Fard. Project Mercury. Don't you know anything.

That's all for fruits.

Just because you don't understand it.

You guys always stand together like that, said Larry.

What's it to you, *kid*, said Fard.

Look, Gomez, these guys got the same jeans and the same lunch boxes.

I don't have the same jeans, said Fard.

Well, they look the same to me, Ugly.

Hey!

See here now, She said, opening the door of our room. Where do you belong, She said to Larry, I think you had better go there and line up, don't you? She eyed Fard.

I've never seen that guy before, he said.

I'm sure that's true, Fard.

Second bell, *in*. She meant business in this room, it was very neat. At the beginning of the year, before they put up the curling crusty stuff we make, a schoolroom is strange. The dotted

16

walls, clock, flag, handwriting chart, new sponge, washed desks. Cork boards, soon to be covered in construction-paper autumn. In a diagram on the wall, two kids kept crouching under their desks, I hadn't had time to study this, but I was already afraid of it. Fard and I were always looking to see what in the room we could put in our scenario of Her.

Now I felt funny sitting next to Fard. Was She going to think our jeans were the same?

Boys and girls, let me remind you that the bell has rung. What is the first thing we do.

I know!

Oh! oh!, straining arm held up by the other, a wig-wag there for hours.

I know, said Linda Johnson, sit down at our desks.

Yes, well, Linda, you should be in your seat by the time the second bell rings.

Ha, Fard looked over at me, Linda Johnson got a question wrong.

Ha ha.

I know, said Julius.

Yes? What is your name, again?

Julius.

And?

Say the pledge to the flag.

That's right, the Pledge of *Allegiance*, that's what we do. Would you like to lead us, Julius?

Ha, Fard, Julius has to get up—

*   *   *

17

When Julius was five or six, he got run over, by an ice-cream truck. Fard knew the story and told me about it. This was the reason Julius had such a bug-eyed look, Fard thought, that the ice-cream truck squished his head a little and made his eyes bug out, but I thought he must have had the bug-eyed look before he was run over, or maybe it perfectly began the very moment Julius turned to see the ice-cream truck upon him, blue and white, its 1940s grille, the Good Humor tune and the way it plays that sounds like blinking. *Aaaaaaaah!*

Man, I said to Fard, imagine what it felt like, getting run over, they're always telling you that, you could get your eye poked out, you could get run over.

Yeah, said Fard, and imagine what it's like to see the words GOOD HUMOR going over you. GOOD HUMOR, *thump thump*, blinkblink blinkblink blink blink blink—

I could think of Julius in no other way, he was *a guy who had been run over*. Everyone else, Kurt and those guys, thought of it in a slightly different way, that Julius was a guy who'd been run over *by an ice-cream truck*, how could you not get out of the way of an ice-cream truck, they go about zero miles an hour. But that never got me about it, Julius was so bug-eyed I was sure there had been a good reason for his not getting out of the way, that it really was a tragedy, I mean, look at Julius.

Gomez used to come up behind Julius in the

18

arcade and do the Good Humor theme, but Julius's mom complained to the Principal.

Beet-red, trembling. I pledge allegiance—to the Flag—of the United States of America. And to the Republic—for which it stands—
   Republicands, whispered Fard.
—one Nation—under God—invisible—with liberty—injustice for all.
   Indivisible, Julius, She said. Do you know what that means?
   She could already have made toast, eggs, Her beautiful breakfast on Julius's face, this hot thing he cast down now and tears ran off onto the floor.
   I thought it meant God was invisible, he said. And he is, too.
   No, who knows what indivisible means? Yes?
   You can't make it go apart?
   Well, yes. It can't be *divided*. It stays together as one. United. Oh, Julius, you may take your seat now. There's—no need to cry, you recited it very—
   Julius stayed up, rubbing at something on the clean desk with a teary finger. Thank you.
   What do we mean when we pledge our *allegiance* to our flag and our country, boys and girls.
   Oh! oh!, now with the finger snapping.
   Yes?
   It means we'll be true to it and fight for it and everything.
   That's right. Why do we need to pledge our

allegiance to our country every day? Hm, She said, I'm going to start calling your names from the roll, I'm still having a little trouble—Eric? Where is Eric? Why do you think we need to say we will help our country, Eric?

Because uh. Because Nazis might come.

Well—

She picked the wrong guy. Fard shut his eyes and slapped his forehead. Eric plays on our street and when ANY AIRPLANE comes over he hits the dirt, doing his strafing noise, screaming *You must die, Schweins*, Ba-dow ba-dow ba-dow, Ktow! Ktow! *Blam!* Eheheheheheheheh! He's ready to go into his movie . . .

No, Nazis aren't going to come now, where did you, that was a war which ended almost twenty years ago.

But they're on tv, said Eric, his eyes enormous, you can see 'em every night. Ba-dow, ba—

That's only to *entertain* you.

Ktow! Ktow! Laughter.

That's enough. Let's calm down now. In the pledge we recite, we pledge our allegiance to our *flag*, and also to our *republic*. What do we mean when we say that? Where is Fard?

I can feel him crease his forehead and perspire.

What do we mean by our *republic*, Fard.

It's the party you're supposed to join.

Gawww, said Linda Johnson.

What? Excuse me, Fard?

And to the Republicands for which it stands.

20

It rhymes, said Fard, our country stands for the Republicands.

Well, that's not exactly right, Fard. It's true that there is a Republican Party, but we have two parties in our country, that's what makes it strong. Who can tell me the name of the other political party? Linda?

She knows *her* name all right.

Democrats?

The Democratic Party.

My father says—

Now, I don't want to get into—listen, you two boys back there? We're talking about what a republic is. It's different from the Republican Party. Now I'm going to tell you what it is. A republic is a country where the people decide, where they *elect* people and send them to Washington, to decide what's good for us. Do you understand that, Fard, do you see the difference. Now, what is a republic.

A republic is where you send people to Washington.

Yes, well, it's where we all, when you grow up you will have the power to make decisions. That's what we call *voting*.

I thought it was *boat*, said Julius.

God, you—, said Linda Johnson.

Now, we don't have kings in our country, do we, let's see, *Gomez*.

Yeah, we do. We don't?

No, and that's one of the things we're going to

21

learn about this year. Our country was founded by some people—

Republicans!

—No, they were people who didn't want to have a king, they wanted to have a republic. What did they want to do, Gomez?

Send people to Washington.

Well, there was no, they wanted all of us, *the people*, to have power, the power to make decisions together, so there would be no bigger—

And now a tremendously frightening, *searing* bell, a red BELL I never noticed on the back wall. Pain hounded by a loping siren somewhere outside, going on and on. Oh my, She said, and froze: Oh, now, this is our drill—duck and cover! Get down under your desks, boys and girls, as shown on this card! She whipped it off the wall, tacks flew. So that's what that is for, I said to Fard. His mouth opened as he squinted at the card She was waving around. It showed a pretty weird way of getting under a table. What if you don't have a table like that, called out Julius. This is a red alert! Duck and cover! Her red nail pointing. Gomez couldn't fit under his desk so he was saying things about all the bottoms pointed at him. The excruciating bell! This is now a yellow alert, boys and girls. Looks more like a brown alert, said Gomez. We scrambled, knocking our seats over, shouting to Her, what to do? I took a last look at Favorite Teacher and the room where I was to have been so happy. I looked at Fard and he at me, all of us crouched there in the earsplitting ringing with our heads in our hands and it stopped. We waited. Now the all-clear came and She said, Oh well, someone must've just dropped a light bulb.

Wait a minute, do you mean something *was* dropped? said Fard, staying under his desk.

That's what they ring when they're going to use the ATOMIC BOMB, said Linda Johnson, gloating with that front-end-of-Studebaker look she had.

Man, said Eric, Ba-dow ba-dow ba-dow! Ktow! Blam!

Boys and girls. You can get up in your seats now. It was just a drill, we've had them before. Haven't you?

Cripe, said Julius, standing up, Teacher, can I go to the bathroom?

Phew! said Gomez. Me too.

Everyone got settled, the door opening and closing on pairs of bathroom travelers. Mrs Dentyne smoothed Her skirt and stole a look at Herself in Her compact. She looked around the room. There beneath us, somewhere, was an insistent, nasty little noise, like termites or guilt. She saw a flesh-colored plug in Simon's ear, his eyes a little shifty. Her eyes widened and a smile began at the corner of Her mouth. I was glad She hadn't seen mine. Simon, plugged into Project Mercury with the latest transistor.

STUFF was everything to him, neat stuff. Simon's stuff was SO neat that they never even took it away

23

from him or anything. Most of his stuff was small, which was important, big stuff like my Chinese pencil last year attracts rage. Simon's stuff lives in the pencil tray under the lid of his desk, it's just TO HAVE. Something to see you through, something from life and fun you can glimpse in the middle of school, just by lifting the lid of your desk in the middle of social studies. Just TO HAVE them I spent my allowance on a bike reflector, a windshield wiper, a short-handled rubber plunger. But Simon's was stuff you would never even think of getting, can't get, pristine billiard chalk, a brilliant blue none of us had ever seen before. He wouldn't waste it on the black top or even try it on the chalkboard. The way it was wrapped, in a tight cube of red and yellow paper, the little Brunswick man. A skull, the size of a rubber ball, carved of gypsum, with a patina of coffee, I imagined the military brother, weird stuff from sick shops in Manila. Miniature calipers, just right for measuring the skull. A heavy little glass from which the dentist takes his mysterious toothpaste onto the high-speed rubber cup. He wasn't selfish, Simon, he'd trade . . . All last year I angled for the gypsum skull. I almost got it in May, for a green cap Luger I found in the oleander on the way to school, but when Simon opened the compartment it was packed solid with dirt, so that deal fell through. You could trade him for Fan Tan or Sen-Sen.

My transistor only got KFI, but it had the same

24

fleshy earphone with pain hook which Simon was wearing right there in the middle of class. The Science Reporters on tv wore them too.

I know some of you are interested in the Project Mercury mission today, and I see some of you have brought radios. We can't get a tv—

Ohhh!, from everybody, all the kids with plugs in their ears jumped—

—at least until later. But I don't mind if you tune in once in a while, as long as you use your earphones.

I looked over at Fard. Isn't She great? You can listen to your *transistor*.

What if your transistor doesn't have an earphone?

Then you'll just have to wait until recess, Julius. Joe?

ME? She hadn't addressed me in weeks, I thought I was safe for the rest of the year.

Can you tell us where the capsule is now.

Fard looked over, he didn't have a transistor. In my ear, *The Associated Press reports from Mission Control that the capsule has just passed over the continent of Australia. And now let me take a moment of your time to ask you about that cup of coffee you're drinking*. She, *She*, was smiling at me and afraid I would mispronounce it, I did. Africa.

Africa, my. They travel so fast, don't they, class.

Yeah, from Australia to Africa in a second, sneered Linda Johnson.

Now, let's—can anyone tell me what they use for fuel in these big rockets, what makes them go.

Oh! oh!

Me me me me me me me.

Nancy.

P. U., said Gomez.

Who said that? Nancy, do you know what powers these big rockets that our astronauts use to get into space.

Fire?

Well, in a way, not—yes? In the back?

Chemicals.

Very good. *Chemicals*. And today I'm going to pass out the Science Book we're going to use this year.

Fard started beating his fists on his desk.

Please don't—I think we'll do a little reading about chemicals. Chemicals have a lot to do with what's going on in space today. Who are our book monitors? If you'll pass these out to the boys and girls.

Fard and I sat with our mouths open, we couldn't believe our fortune, Project Mercury, and a real SCIENCE BOOK, on the same day. And HER. Now we really were going to study Science, not just make water change color, or stick something in something and *wait around*.

Piece of wood, stick of butter, chunk of coal, bolt of cloth, bottle of ink, of soda pop, quart of milk with a paper hat, sack of sugar, sheet of copper, cube of ice—in his shrine of the rear endpaper a MODERN CHEMIST grinned fiendishly, almost, across the centuries of pages at the Humble Alchemist of

26

the front endpaper, kindly among his medieval skulls, my copy'd been dropped in something last year, it was a little wavy. Should I write that in the CONDITION column of the stamp, *wavy?* All the pictures looked like they were from the peculiar watercolor box of HERBERT S. ZIM, Editor-in-Chief of the Golden Nature Guides of *Racine, Wisconsin,* my King.

Science Heaven. Fard was swinging his big turned-up jeans under his desk.

My, it's hot, isn't it, boys and girls.

She walked around, happy as we that we all had our Science books, Her finger on the page She was going to read out. Her forehead was large and high. That brow could worry and it was damp under Her makeup. She wore a red dress with a white collar and white piping, the ends of the short sleeves on Her pretty arms were piped, the pockets, over Her breasts, were piped, I was piped. I could see two wet places under Her arms when She lifted up the Science book. She sat on top of an empty desk in the first row, crossed Her legs and swished Her ponytail in a downbeat to Her pleasant reading:

Everything in the world is made of only one hundred different substances.

Man, said Fard, is this great.

These hundred materials are called the elements.

Looking at Her reading to us, perspiring a little, that forehead shiny, I could see it, I saw it all in front of me, of course, I had spent all my time at the kitchen table with Dad only with COMPOUNDS. Why had he kept the ELEMENTS from me?

The building blocks of our world, She's saying . . .

Till now, I thought, they were colored water, gum, tv, school clay, dirt, plaster of Paris, rubber, plastic, food, heat, bugs, wood, nails, tears, and stuff you can bend. JOE LAKE'S PERIODIC TABLE OF THE ATOMS. Beakers might be an element.

I was waiting for Dad to come home from the Lab. I had certain WAYS to look out of different windows, the window in the front door was covered with gauze stretched tight and pleated. My eyes only cleared the bottom, so I pushed the gauze aside with my nose, sometimes with my tongue. The gauze tasted tart, of cool dust. Soon one of the pairs of headlights in the 'rain' would turn and it would be the brown car. Everyone came home at the same time. They ate, they did things, they all went to bed at exactly the same time. By this you shall know: suburb. Dad had promised me a beaker from the Lab. He thought I wanted it to experiment with, but mostly it was just TO HAVE.

I had visions of this beaker. From it I would extrapolate a complete laboratory, I would stir colored water and invent things with dials on them; Anita would think I was a Genius and play with me all the time.

Dad was troubled by the beaker. He had never brought home so much as a pencil from work, it was theft, THEFT I'm telling you. He kept not bringing the beaker, kept forgetting it. Since he was late of course I thought tonight would be the night, I was too excited even for the cornucopia of the dark, even though Julie had carrot sticks and Bugs Bunny in there.

I went to sit behind the sofa, I had waited so long I was forgetting what it was all about, finally he came in, it was one of those moments when you wait for someone and you get to thinking that THEIR only thoughts are of YOU, which is usually far from true. Dad had been thinking about taxes, the rain on the road, Castro, the unimportant idea he was late; and the most unimportant thing in the world, me and my goddamn beaker, he had forgotten to think about. *I'm home!*, absently. He had a way of looking at you as though he was embarrassed about it, of course, but he'd forgotten about you entirely. Because he was a Scientist.

Dinner was oppressive because of the hour. Instead of relaxing because he was late, Dad would do something dramatic and irritating, turning on all the lights in the kitchen while we ate. All this light made our food look PITIFUL, like the fatty washed-out photographs in the cookbook. Julie began to cry. We had no checked tablecloth but somehow one appeared beneath dinners like this, under the harsh frontier-style light, a little brown chop, wet and over-familiar, soft lima beans of bathroom green, rice with a pat of butter that would not melt.

I wished we had other plates, the brown coaching

29

scene did nothing for our usual chop. The overhead light made me feel funny, Mom and Dad talked on and on about one of those things nobody understands. Zoning. Dad was getting hot under the collar, *I'm telling you honey it is zoned*. These were tenacious arguers, Julie and I were used to it. Neither was above cheating in order to win, either. The dinner trailed off, ragged, no feeling you'd eaten anything.

Dad was so irritated he did the dishes himself. To torture himself more he turned on the loud fan over the stove. There was something weird about Dad doing the dishes, the thought of it always upset us. He didn't put them in the rack the RIGHT WAY.

Wait a minute, he said, drying his hands.

I thought he had something about zoning he was going to get Mom with in the living room, but he came back with a bag.

Wellsir, here's something that might interest you.

He brought out glass rods, flasks, three or four vials of chemicals.

This one's yours.

And with a flourish handed me the tiniest beaker I could have imagined. The result of his inner struggle. 5 ml. So much for my idea of copping Florence and Erlenmeyer flasks from him too, you wouldn't have been able to see them.

I watched as he arranged the really really Scientific glass and chemicals, rods, papers, all of which would go back to the Lab tomorrow. If only Fard were here. On the plastic tablecloth he laid it all

30

out plain and clean as Herbert S. Zim. He put on his lab coat. He gives me good lessons in this, I thought, and I never really listen. I just watch him in his lab coat, admire his fingernails, his wire hair and his pipe. Now we were going to experiment.

Science surrounded us, not Nature. Science gave life, made the news, made our food and sterilized it. Science gave us a big orange shot and made us well, Science met with the President every day.

Frank Armbruster has a chemistry set, I said.

Chemistry sets are dangerous, son. I don't think it's proper. Why just last week in Anaheim, a boy blew up his entire parents.

Dad's movements were sure. He raised a flask and a test tube up to the light, mixed them. The confused brooding, then change of colors. He was paradigmatic of Science. We weren't in the kitchen anymore but in a huge pristine Lab, long white benches with retorts, burners, condensers and big beakers. Lots of guys with the same glasses and coats and furrowed brows. Dad was at the front, on the altar of the Lab, he was bigger and more Scientific than all of them. Actually this was the night he set his hair on fire, but he did all that for me because he didn't want to play catch.

—these elements are mixed and combined in various ways, some of which make them very different from the way they started out, She said. For example, if we put together two very common gases—

Gases, exploded Gomez.

31

—things you can't ordinarily see, She said, hydrogen and oxygen, makes—

Farts, said Gomez.

Who—*makes water*, She said.

Makes water! Gomez collapsed into the fat cave of his arms.

And, She went on, your father's cup of coffee, that he had for breakfast this morning? Keep reading from here, Nancy.

—is a mixture of chlorine, a poisonous green gas, and sodium, a poisonous metal.

No, I don't think that's where we are . . .

But put 'em together and hmm boy. Guy it's hot, I'm drifting . . . without living carbon we could not live, for every living thing alive living in our bodies is made partly by living carbon, carbon is a living part of the life-giving material of every living thing. The boy in the left-hand picture is using the limewater test to find out whether he has any carbon dioxide in his breath. In my ear, that cup of coffee you're . . . Project Mercury, Madagascar . . . You cannot do anything interesting with nitrogen. A gas can be poured like water. Sulphuric acid is the most important sulphur compound, it has many uses, among them making ammunition—

Ktow! Ktow! said Eric, hoarsely.

—*and fertilizer*, Her hand to Her lips, getting gasoline from petroleum and jamming radio broadcasts from Soviet Russia—

I need air, I said to Fard.

—sometimes it is called the King of Chemicals,

though nitric acid is important too. Everyone is smiling in this picture. What do they know that you don't? Compounds of nitrogen are used in explosives—

Blam! *BLAM!*

—Shhh, which play an important part in building roads, mining, and the everyday work of waging war. Knowing what you have just learned—

Fard? My head is whirling.

Why don't ya—

—about the elements, *boys*, why do you think red socks keep your feet warmer than black socks? *Joe?*

Oh no. Ah—

Let's pay attention. Julius.

What looks hot *is* hot?

Gases have no shape of their own, Julius. It is ridiculous to think of making air into a model of a little animal. Do you want some $C_{12}H_{22}$ on your oatmeal? Shall I put some ice in your H_2O? Is there enough NaCl on your egg? Suppose your mother asked you these questions at breakfast this morning? What would you have answered? *Joe?*

But my father did ask me those questions this morning.

Oh, ha ha, he did? Wh—

My Dad's a Scientist.

He is not, said Nunzio.

Mine too, mine too.

Now—

The room was very hot and the impending year of

33

Science seemed noisy and was making me feel sick. I was trying to answer Favorite Teacher and get Fard to tell Her my Dad is a Scientist, and turned just in time to see Simon open his mouth and with one convulsion cover his desk, and only his desk, with a collection of recognizable breakfast suggestions. Nancy Hoffman started screaming, and everyone pushed their desks away from Simon's and at the height of revulsion the bell, the blessed bell of recess.

Dust, the Big Tree, the fence, another year of this, I thought. There was still no tv to watch Project Mercury so we had to get out there and do it ourselves.

The important thing about outer space was to have CONTROL, a panel, a headset. When you watch space stuff, that is what you see, GUYS IN CONTROL. You get sofa feeling from it.

Nunzio walked like the creep he was toward the Big Tree, insisting that was Canaveral today. I waited for Fard and we headed for the Big Tree along with Kurt and Julius.

Why do you think they call this the Big Tree, said Fard, I mean, look around.

What do you want to call it, said Kurt.

I was thinking of, The Tree.

Fard was still excited by our Science book, he brought it out with him. This is great, he said, thumbing through.

Nunzio didn't like Fard, and made a face. Did you guys see the launch this morning?

34

There was a hold at T minus four minutes, said Julius, but they didn't say what for.

Telemetry, jerk, said Nunzio. Don't you watch CBS?

What if your tv doesn't get CBS, said Julius, red now, his eyes filling with tears.

Get a new one. Stupid.

Nunzio had this idea that there was one guy in charge of the mission and he had a guy sitting on either side of him. If you weren't one of those guys you had to be in the capsule or be a tv guy. Nunzio got a real headset from the telephone company and that made him always in charge, the rest of us had to hold the sprayers from bottles up to our heads.

Where are we gonna take it from?

T minus ten minutes, said Nunzio.

Nunzio picked me and Julius to work with him during the countdown. He put Kurt in the capsule, he liked being there, plus he had a Virgil Grissom haircut. He settled against the Big Tree.

We were at T minus three minutes when two horses came into the launch area. The idea behind the girls' encounters with us was: WE ARE HORSES. Anita and her friend BECAME HORSES. They cantered over and sniffed our game mistrustfully.

I was standing at the edge of the black top last year, by the drinking fountain, with Billy Williamson, when Anita galloped up to tell me something Scientific. She and I could really sling around a lot of electrochemical language. She delivered it in her self-satisfied way, whinnied,

35

and raced off. Williamson looked at me and shook his head, he spoke like a parent, *Joe, you're going to marry a Scientist*. What cut, dry vision he had.

Would you stupid horses get out of here, said Nunzio, we're getting ready to fire the rocket.

She snorted and pawed the ground, looked at her friend, said something sotto-voce in English, then in horse, for our benefit, then they galloped away across the field.

What'm I, Nunzio, said Fard. Standing there.

Nunzio looked at Fard and his eyes and teeth grew particularly cold and hard. Where's your sprayer.

I don't have one, said Fard, my mom still won't give it to me.

Tell you what, Fard, said Nunzio with real contempt, you know what you can do on this mission? What.

You can be in charge of *laundry*.

Julius laughed but turned red the moment he got a load of Fard's face. Nunzio stood with his hands on his hips, determined and cruel, like a determined cruel guy on tv, watching Fard stumble away, starting to cry. He walked away from us fast and started talking into his Science book, explaining his hurt to it as he got smaller.

We were in a kind of thrall to Nunzio. He was in charge of the launch because he had a real telephone company headset, and when he fixed you with his helplessly cruel eyes you had to do exactly what he wanted. Fard was crying, across the field, I could see that. Fard crying. But we were bursting here with our sprayers, the need to get the launch started, Julius's watch ticking. I betrayed Fard.

How can you not get a sprayer, said Nunzio, staring across the field at Fard with hate.

Well, I said, Fard is kind of poor.

How can you not get a sprayer, Nunzio said again. Nunzio had a swimming pool and every toy as seen on tv.

T minus one minute and counting, said Julius, hiding from all this behind his watch.

Now I wanted out, immediately, and said, which I did not understand as it tumbled from me, that although they were very poor, Fard's dad *made* him a Think-A-Tron for Christmas. Yes he did. It had the same cards with questions, WHERE ARE THE PYRAMIDS?, and you put the card into the Think-A-Tron and it made a grinding noise and lit up B, GIZEH EGYPT, on the front, Fard's dad reproduced it exactly, yes, he had ways of molding plastic and EVERYTHING.

Liar, said Nunzio.

But I really couldn't stand for Fard to be poor anymore, I didn't want his mom to do laundry all the time, I didn't want them to Broast. I wanted Fard's dusty folded dad to be handy, instead of the slow-moving driver of a carroty truck. I wanted us all in Project Mercury together, and that is what the President wanted.

37

Okay, said Nunzio, let's get—
The bell.

Guess what they did, I said to Dad last week,
they parked this dumb-looking TRAILER in front of
school and some kids have to go in there for *Bible
Class*, ha ha ha.

Like you.

What!

Maybe they forgot to sign me up, I kept thinking,
but we were no sooner in our seats than She looked
up at the clock and got out a list, which I was gosh
dang *on*.

These students will please get up quietly from
your desks now and go to the Trailer of Prayer.

At least Kurt and Fard were going too. Along
with a lot of jerks like Breen and Buzza. We walked
along the black top in the heat.

It's two hours till lunch, said Fard, in a doom
laden voice.

This could be very bad, I said. The school seemed
dry and bleak, I thought of the clock on the wall
back in class, its red sweep and enemy face.

The Trailer of Prayer had a suffocated look. I
pointed out to Fard it was on the exact spot where
Murry sold us slushes after school.

Maybe we can complain, said Fard.

Then Breen got up on the step and KNOCKED.

What an idiot you are, said Kurt. If they send
you somewhere and nothing seems to be going on,
you just leave, man, you don't *ask for class* when

38

you don't have to have it.

The door was opened by a cranky lady.

Come in, children.

Oh man, said Fard, bad news. *Children*.

Although there were windows, they were covered with pieces of vinyl. Maybe we were going to see a movie, I thought, that would be OK, a few weeks of Bible audio-visual. A snap, like Health. But there was no movie, the Trailer of Prayer was dark and airless to preserve the religious awfulness of this lady, who turned out to be an extra special horror sarcophagus mummy.

Why does the Bible, I thought, this ugly black book, pursue me? I felt trapped with my Aunt in one of her surprising summer vacation Bible lessons which could happen at any moment, any time she could back you into a corner, behind the refrigerator was her favorite. *Let's talk about our Lord.* Jesus had something to do with not being bad. So? I am not bad, I thought, the only thing that makes me want to do stupendous, horrible things is when Aunt, and Uncle, and Grandma Lake, and Mrs Neighbor, that weird kid from Clark Street, my tap-dancing teacher, and even Mom, after Grandma Lake told her I hadn't had a bowel movement, chase me with the Bible and try to tell me what's in it. Then I want to do things, grim reminders and mute testimony out of books, things with swords and fire and the inky blackness of the sea. What they're always yelling at me from the distance is that only Jesus knows if you're bad or not. But that is like the vacuum cleaner salesman who brings his own bag of dirt to suck up.

The lady acted like we'd been there before. She started right in asking us questions about all those tragic people with plausible sandals on.

I don't get this, I said to Fard, I thought they were supposed to teach you stuff and *then* ask you questions.

Fard looked intent on this junk and nervous. Don't you know this, he said, raising his hand to answer *Peter, called Simon Peter*.

I felt like my Aunt had her hand in my guts. This was really a *class* and you were supposed to know already all about these people. They never really told us about them at Sunday School, which I got thrown out of, they just showed us a bunch of pictures, come to think of it *they* thought we already knew about them too, what is this.

This is the stuff you can never figure out, I thought, the stuff that stays on your mind dial forever, I bet you dream about it when you're twenty years old. What was I doing in this trailer, it makes you think they're trying to get rid of you. Maybe they are. They're always putting you in places they don't know about, how horrible they are. Or maybe they want you to know what it would feel like if they did get rid of you.

One day during vacation Mom said, you're going to spend the day with *Malik*. He was the son of one of

40

the Scientists at the Lab. They were from ANOTHER
COUNTRY.

What for.

You just are. You'll be visiting Malik at his
nursery school.

Nursery school! You said they were bad.

A strange atmosphere always blew up when you
said nursery school, it was something Mom and
Dad didn't approve of, their endless list, thanks a
lot, training wheels, the Three Stooges, Kool-Aid,
Robot Commando.

What do you do at nursery school, anyway, I said.

Just a lot of stupid things.

Great.

The place, Peter Pan Nursery School, was a
house, the front and back yard were filled up with
play equipment. It smelled like Sunday School,
there were a lot of kids who stared, it was run
by two ladies in glasses, the whole scene STUNK,
it was way too organized. This was *vacation*.
Mom said goodbye and got out of there, you can
always tell when they're leaving you someplace
stupid, they PEEL OUT. The ladies made us line
up, boys on one side of the main room, girls on
the other, we stood there solemnly, like church,
now I was really starting to hate it. They put
on a scratchy record, real loud, everyone's eye-
lids twitched because they feared the music was
going to be WAY TOO LOUD from the scratches.
Up on the ceiling there was a big public address
system horn, the kind that grows a mouth and

stretches itself out real long while it yells something frantic at you. A cousin of the steam whistle with the fingers. Malik stared straight ahead with his eyes bugging out. Now, flute playing, which in fact you could barely hear because of the scratches, flute music, up and down. Everyone stared up at the horn. I don't like flute music, it just goes up and down and makes you feel like you're wearing a dress. I thought about the green part of my mind dial.

I always thought about my mind when I went to bed. I thought of it as a dial, like the spinner in board games. Julie liked Forest Friends and Dad hated it more than anything, it actually made him purple some nights. You had counters and the board was a trail printed winding through the forest, whoever gets to the end first was the winner. What's wrong with that? It was for little kids so nothing bad could happen, you didn't have to go back or anything that might make you cry, it was just like Dad said, *spin and move, spin and move, spin and move, I'm going out of my mind with this spin and move.* Even POPCORN didn't make him feel better about it.

You can choose the pictures on your mind dial. When you have complicated thoughts or crappy feelings you stop at that picture, then you can go into it and look around, if you want, probably not if it's a picture of your Aunt and her Bible as I was just thinking.

42

You have to make sure your mind dial is spinning before you go to sleep, so you'll dream.

But when I was listening to the stupid flute music I was made to think uncontrollably of the squirrels from Forest Friends, it was bugging me and I felt way too leafy—*then* flute music always tries to sound somber and spooky and that's when it really gets DUMB.

All the kids looking up at the horn, do they do this every day, then a guy's voice came on, one of those guys who's afraid to talk in a normal voice, I thought, he needs lessons from Bill Welch. I started to feel creepy as when the dentist played his records to me, the ones with all the talking, *Barry Goldwater Reads Favorite Tales for Children.*

These are the pipes of Peter Pan, the voice said. Then everybody stuck their hands up in some kind of salute, so they did do it every day, they promised to be playmates forever and hold fast to the spirit of youth, let years to come do what they may. Girls across the room stared at me like I was crazy because I didn't know the pledge or whatever it was. I thought this was like the Army Fard and I watched in the movies, the pilots always got their missions from big horns.

The lady excused the girls and I started to go with them, just because it was a big group of people moving, not because I'm a GIRL. Malik grabbed my shoulder, he acted terrified by my mistakes, what, are we going to be shot. Everyone laughed and the lady said *just girls, that's right, Malik*. Everyone stared at me for the rest of the day because I didn't know the Peter Pan ceremonies, I'd never been in a place where girls did things differently from boys, how could it be that you had to hold a paint brush opposite to the way we did in school?

Kool-Aid (see?) and cookies under a vine full of BEES so you couldn't relax, they wanted it that way. Malik's sister ate with the girls and I felt sorry for him, he couldn't talk to his own sister if he wanted to. Everybody here was learning such weird stuff, I told Fard later, I was frightened of them, their scary school in the middle of summer, they thought everything they were doing was absolutely right.

How was your day, Joe boy? Mom said when she picked me up.

I'm never going THERE again, I'll tell you that.

At our half-desks half-pews in the Trailer of Prayer we had Supermarket Bibles open in front of us. I didn't like to think of someone making this into the Trailer of Prayer, who knew what the whole history of it was anyway, maybe it was slept in for years by a farting family who ate baloney sandwiches out of waxed paper bags and the mayonnaise was all hot. It had the same resinous staleness as Fard's trailer, when we came in I thought of his sister, maybe I could sit here for two months and think about her bathing suit. The smell made you think of blops of glue that must be behind the walls, trailers have to be light so that Jesus and God can hurl them around Mississippi and Florida, that's where all the sinful people live, in trailer parks.

And here were the goddamn pictures again. Is it really okay, I whispered to Fard, to draw people in the Bible like they do in comic books?

Well, I guess it's blasphemy.

The bearded guys with the blue-black hair, their *plausible sandals*, which reminded me of the fanatical footwear of my Aunt, looked like they were getting strafed in *Sgt Rock*, these pictures, how were you supposed to tell one guy from another, and now she was really asking me a Bible question. I knew so little of the thing I couldn't even guess what she might have said, I was going to say the blue-black guys were the Pharisees, they looked like they were getting strafed Pharisaically, *die, Schweins*. Oh *I don't know*, why not admit it, she was already drawing herself up, stiffening, drying up, freezing me below the point of molecular motion, she saw I was evil and would have to be given scriptural homework and here this was going to be another hell. She turned, grinding her neck gear like a lighthouse, to Kurt, *and he knew it*.

How did you know that? I asked him on the way back to class.

It's in the Bible. I just know it.

A big hand came down and moved me like a checker a thousand miles from my friends. I thought I knew them. This bible mummy gave them another reason to think I was weird, thanks a lot. I couldn't make jokes with them about the Bible because they knew this stuff, they seemed perplexed that I couldn't answer simple questions about blue-black people in the Bible. Kurt and Fard didn't even go to the same church, and they were *in* on it.

Oh! You students are back just in time for arithmetic.

Can you beat this, I said to Fard, they send you to a thing like that and it doesn't even get you out of arithmetic.

Yeah, he said, I wonder what we *missed*.

When they first made us count and add they gave us pieces of black construction paper and some disks. Some of us got cardboard disks and some of us got SLUGS from electrical junction boxes, slugs were like money and they were very neat to have, as stuff, like ball bearings, which showed marbles to be the dumb crap that they are. You were supposed to go into houses being built and pull the slugs out yourself, it was DANGEROUS and you could get in TROUBLE. Slugs were so neat that a lot of people stole them from the arithmetic cups, not me of course, but I worried about them and tried to buy some at the hardware store. Our second-grade teacher couldn't understand where they went, she didn't know what they MEANT. To her they were just metal disks someone'd given her, the dingbat.

46

She called out a number and you had to arrange
the correct number of disks in the right pattern,
like dice. They're always doing stuff like that to
you. But that was the last correct thing I ever did
in arithmetic. Things got so bad that I knew I had
no idea how bad they were.

My lack of understanding of arithmetic is SO BIG
that at times I think I'm doing everything right.
But I can't even give an example of right *what*. Half
the bad dreams I've had since I was born are about
arithmetic, people in offices shaking their heads.

Last month Favorite Teacher handed out workbooks and said to us, *when you're done with these
exercises you can go out on the field and have free
play*. Free play is when you have fun instead of
playing kickball.

This is easy is a dumb thing to say, you should
always be afraid when you hear yourself saying
that, it automatically means that what you're
doing is impossible and you are blowing it and
people are going to ridicule you and EVERYTHING.
I was so messed up, the problems on those three
pages looked short. I knew what I was doing, I
whipped through them, I couldn't believe it, my
heart was singing.

I put my workbook on Her desk, She really
had little inkling of the boundless scope of my
difficulties as yet. She knew I wasn't great at
arithmetic but She didn't know the picture I had
of it on my mind dial, the crater of a volcano,

47

confusion, lumps of things, muffled cries from all the people who always told me how to do it. Mom, Dad, Julie, Fard, lost to me. Then sometimes it was the inky blackness of the ocean, where the wreck of my arithmetic lay, a grim reminder there was something wrong with me. But I listened to my heart sing that I could solve these problems, that *understanding had come to me in the night*. Now I was going to be like everybody else.

I went out to the field, which seemed beautiful and great, like a blue-green day from a movie. A couple of kids were swinging and Billy Williamson was playing football with the Czechoslovakian kid from the other room, the kid who called everyone FARMERS when he meant FARTERS. Or so we thought. I got on a swing. Man, I thought, all you have to do is fill in those *blanks*, with those little *answers*, now that I understood how, I'd hardly done any thinking, maybe that was the secret. Then you come out here and things are really neat. I felt like I'd made friends with the world. After a while more people came out, and She came with them, Her hair in the breeze and the pile of workbooks in Her arms. Her eyes like the little flowers that grew next to the crab grass on the dirt of the field. I was swinging and swinging, talking to Fard now he'd come out, no one had said anything to me, ME, about being one of the first guys out here, I was so stupid I didn't even think about that myself, the IMPOSSIBILITY. She hadn't even given me a look when I put the workbook on

Her desk. She called to a couple of people, minor corrections, they stayed with Her for a minute and then went back to their free play, She must have looked at my book already. Then I heard Her sweet voice calling *Joe? Joe Lake? Come over here please*, and there was a new tone to it, PAIN is what it was. Something a little bad started to happen way at the back of my head, it wasn't exactly arithmetic sickness, but maybe She—

Cripe, here was my workbook open in Her flowered lap.

Joe, what were you thinking when you handed these in?

She didn't sound mad, She hardly ever gets mad, and when She does it's kind of funny, Her forehead gets larger, and it de-powders, and She gets a really weird pink color as Fard has pointed out. He makes a Science of Her skin. But Her voice was hurt, She couldn't believe I would turn in the wrong answers, not just wrong answers but answers *which revealed* I did not read the directions, that I have understood nothing since 1958. Answers which showed I hadn't listened to a thing She said all day, all week, all month.

All year. I couldn't believe it either, how come these small problems were so difficult, why can't you just answer them the way they look like they ought to be answered, small easy ANSWERS in just a second? Now the crappy feelings started, not only was I really dumb but I'd hurt Her, betrayed Her, She would have to be RESCUED by Fard, from me.

Without being mean She told me to get back to work, which I couldn't do, because the way I did the problems was the only way I could do them. And I could only do them once.

What are you supposed to say, I asked Fard on the way home, it doesn't do any good to tell them you wouldn't even be able to get THOSE answers again. How could my brain do this to me and why, I thought. Magic and hope are crap.

Last year wasn't good. Fard and I had this really mean teacher, and she was *really* mean about arithmetic. She made you stay after school a lot when you messed up but she didn't HELP you, she just sat at her desk doing her own stupid stuff. Once in a while she'd look up and shake her head or say something really mean, *I don't know why some children can't settle down and learn*. Fard and I used to talk about her the whole way home.

Some children, said Fard, I hate it when they say that, it makes me feel like I'm going to have diarrhea.

Are you?

Nah. She doesn't even tell you what you did wrong.

I know, I said, like if you sit there long enough you'll just start writing the right answers. That's what's wrong with arithmetic, that's why I hate it and it shouldn't exist: there's no why. In history and spelling and all that stuff you can always ask why, WHY did George Washington start fighting England, you ask them and they tell you.

Yeah, said Fard, cause the English guys were being *really mean*.

Or you ask them, why does it rain.

Because the clouds get full, said Fard.

Anita wouldn't have liked that answer, I thought, but in arithmetic, I said, taking in a huge breath, you can't ask why, you just have to believe them when they tell you stuff, they get *mad* when you ask, they're telling you *this is how you add*, and you say well, why is that number on the bottom, and then Mrs Plank looks at you like she wants you DEAD, NOW.

Yeah, said Fard. Tears came to his eyes.

And then her mouth gets real sour and she puts down the chalk and says *I don't know, some boys just don't want to learn*. But you're TRYING to learn—

Damn it, said Fard boldly—

—and you're asking questions about it, isn't that what you're, but questions have nothing to *do* with arithmetic. And they never even said you're *not* supposed to ask questions, they *said* ask, but then they go crazy when you do.

Yeah! said Fard.

Mrs Plank hates me because I'm not *moving forward* in arithmetic, I said, that's what she said to my own mother. I'm sitting in my seat wondering why why why all the time, I *am*. And there are no answers in arithmetic either, in addition to no whys. The only answer they can give you is IT JUST IS, THAT'S WHY and then they go back to yelling this

51

junk at you they can't explain, that's it, I said to Fard, *they can't explain it* so they yell at us.

You know what I'm going to do, said Fard, I'm going to start saying that when she asks me about my answers. It just is, that's all. You old bag.

We laughed hard and bitterly at this, crossing the Big Street. Fard started talking about how his brother let him drive their pickup truck back and forth in the alley. Fard was the only kid who was interested in driving, everyone else was still riding bikes, I still had my tractor, my *spazz tractor* as Nunzio called it, thanks a lot, and my Small-Wheeled Bike, something Dad embarrassed me by asking for at the hardware store, *we want to get a small-wheeled bike for this boy*. There were some people who had full-sized bikes, they went bobbing up and down, they had bruises and bandages and scabs.

What do you want to drive the truck for, I said, I bet you could get hurt really bad and it's AGAINST THE LAW.

Fard looked like I'd insulted him for a sec, but then he said, when I can drive I'm going to run over Mrs Plank in the dark.

Great idea, I said. Man, if you ran her down in the inky blackness, you could leave her body there as mute testimony, a grim reminder to other teachers.

Yeah, said Fard.

He was right too, *I'd* like to run her over, she's still alive. Still alive.

Even Linda Johnson couldn't make herself into Plank's pet, which shows you how mean she was.

I wasn't like Simon and the toy kids, I didn't have to have something to play with right inside my desk to fondle at the first sign of trouble, but Mrs Plank made me feel so bad I decided to bring a special pencil for arithmetic. If I dedicated a pencil I really loved to arithmetic, it would help me, in the inky blackness of arithmetic the pencil would stand as mute testimony to my efforts, I was getting these words from the book I read every night in bed, *The Illustrated Book of the Sea*, which seemed to be more about arithmetic than mollusks. And if I died of arithmetic my special pencil would be a grim reminder to Mrs Plank and Mom and Dad of my wasted life.

One night in Los Angeles we went into Chinatown and Dad bought me a big jade-green pencil, at the end was a wooden Chinese guy with a purple tassel instead of a pigtail. I had to choose between the pencil or clam shells that opened up and let out streamers and flags if you put them in water. The Chinese pencil wouldn't fit in my pencil box so I had to carry it to school in the same hand with my lunch box, for all to see, it was the kind of thing Gomez would really get you for.

In Mrs Plank's class arithmetic came when lunch seemed even further away than it did when you got to school. Fard used to call that *9,000 o'clock*. But I was less afraid when she said *take out your arithmetic notebooks, children*, I felt a

53

little hopeful because I had my Chinese pencil. She gave us page numbers and we started to work, they were adding so I could do some of them. I felt cheerful because the tassel of the Chinese guy's head swung back and forth merrily when I wrote my answers. Also, Chinese guys invented the abacus. I made hatchmarks in the margin just to watch the tassel flip left, right, and then I saw Mrs Plank with her sour mouth, *the world's most curved thing*. She stopped at the top of our row and folded her arms and looked right at the tassel.

You can't use a pencil like that for arithmetic, Joe.

That's all she said but huge tears leaped out of my face and splashed all over my desk. I wished I could think of Mrs Plank as Teacher, as most people called her, then her name wouldn't have kept me awake at night. I whipped out a yellow pencil, you had to do stuff very fast for Plank. I hated the way it looked, and having to read those dumb words over and over, after every problem, BONDED LEAD, M, **CERES** BY MUSGRAVE PENCIL CO. SHELBYVILLE, TENN. 909 N⁰ 3. The word *Shelbyville* started to drive me crazy, *Shelby*, the sound *Shelby*. So Plank was happy now, she hadn't helped me with my arithmetic but she'd told me how I *couldn't* do it, with a *Chinese pencil*. When she turned away I put my head down on my workbook and cried, utterly silent, mute testimony, maybe I could soak the thing to death. Julius turned around and looked at me.

Guy, Joe, he said, just because you're lousy at arithmetic.

But Julius didn't care if I got arithmetic or not, he just hated crying, because crying made *him* cry. These surprising big tears were hot and they smelled like chalk, they were just like the tears they had for you up at the blackboard. Mrs Plank made you go up there exactly because she knew you didn't know how to work some kind of problem. She'd let you boil away to almost nothing up there, she did it to Kurt a lot. *Some boys just don't want to learn, I s'pose*, she said, *you can go back to your seat now, I guess*. Did she ever show him how to do it, NO, she got Linda Johnson to do it which meant nobody cared, they'd never do anything *her* way.

While I had my head down on my desk there because everything STUNK and there was no way to do arithmetic, they took your very pencil, I uttered my word. I have a word I discovered which I said to myself, violently, when I was very very angry. I discovered it by its sound, it doesn't mean anything, but it sounds horrifying and saying it, even silently, calms you by getting rid of crappy feelings AND stirs them up so you can take a bath in them. Sometimes when I was mad at home I wrote it on a piece of paper and then tore it up in a rage and threw it away, that was as exciting as saying it. So I said it there, in Mrs Plank's class, silently and violently. I'd never had to use it in school before.

*　　*　　*

55

But I used it a lot in my room, that's where Dad 'helped' me with my arithmetic, by driving himself crazy with my stupidity. Since he made the blackboard in my room he thought that gave him the right to make me do arithmetic at it, which just about wrecked my whole room, like filling it with deadly gas. The feeling of looking at his unendurably precise writing on my blackboard, twenty-five problems, the sound of the door shutting, *call me when you've finished*. I thought about dying, or setting fire to the house, my own things being used against me. I could take the sponge and erase all the problems, if only I could come up with some explanation. *I don't know who that was*. Or, *what problems*. His neat writing on my blackboard made everything less colorful, the world was drab and oppressive, the light in the ceiling glared out, really mean, and I started hating the cowboys on the walls and curtains, watercolor cowboys who roped nothing. Even if I listened as closely as I could Dad's words didn't make any sense, there were the whys, I wanted to ask at least one why between *each* of his words. But why gets you nowhere. I made a stab at the threes and fives, I could do those once in a while, but then he was coming back IN, with these QUESTIONS, *Don't you understand this, WHY don't you understand this, what seems to be the problem here, looks like we're going to have to shoot you down on this one*. Then his OKAY, meaning *I'm going to explain to you everything from the beginning,* starting from the Phoenicians or how you make fire probably.

Whatever they're doing to you, a moment eventually comes when the torture is over and you're crushed when they leave. Even though Dad sent the armies of arithmetic against me, I was scared and even more alone when it was over. *Wellsir, let's call it quits there*, and he went out to tell Mom how dumb I was.

Last month, after that farce of free play, Mrs Dentyne sent a note to Mom and Dad, and we had to go to Los Angeles, to a big bookstore. Mom and Dad wouldn't talk to us, they huddled in one aisle while Julie and I roamed around. She required a book about ducks, while I found *Mr Atom*, full of powerful grey pictures of electrons. There were pictures of atomic reactors, atomic power plants, atomic bombs.

What is that about, said Julie.

Everything atomic is good, I said.

I almost threw up when I saw what Mom and Dad were buying, arithmetic books. One was the same stupid book we had at school, without the decency to look worn out, so that gave me the sickness right there. The other was the most horrible thick-looking stupid workbook, *Arithmetic Town*.

God, I hate the word workbook, said Fard only last week.

The cover was red and black and there was

a picture of one of those towns that have two-story schools, parents in cars waving, kids heading into the school looking expectant, not like they're going to get kicked in the ass, a few stylized dogs that wouldn't bite, clean roads, a stoplight. The flag. Houses with TREES across the street from the school, everything looked like fall and spring at the same time. These two stupid books cost a lot of money. Dad wasn't too happy about getting the duck book and *Mr Atom* too.

You have to make it a pleasant experience for them, said Mom in the car.

Pfft! The purpose of this trip, said Dad, is *arithmetic*.

I said, I feel—

Not here!

Now came the end of afternoon as I had known it. I had to leave Fard at my own door every day to spend two hours with, *in*, *Arithmetic Town*. It was enormously, endlessly thick, the pages were so thin you could tear out fistfuls at a time and it would make no difference.

It's supposed to last your whole life I guess, said Fard. I hope you get better soon.

God, this was bad, and *Arithmetic Town* was going to wreck Saturdays too. At least I would be safe from it early in the morning when I got up to be with Laurel & Hardy. They hated arithmetic. While Mom and Dad slept like beasts in their warm bed they couldn't make you do anything.

But Saturday had been OK until this happened, I was allowed to go to the variety store, sometimes I'd take Julie. If I saved my allowance three weeks in a row I could afford to get a model.

The models were all the way in the back of the variety store, past the baseball stuff I never looked at, past the games, skates, cowboy hats and past the stupid craft kits, make your own Indian moccasin, well, whoop-de-do.

That's the kind of thing, I said to Fard, that your *Aunt* gives you, if she doesn't send you a dumb book about God.

Mine too, he said eventually, to be on my side.

I didn't know that Fard knew all about God. But he was way too neat of a guy ever to make a craft kit, could you imagine your best friend wearing grey felt with junk glued on it?

The model shelf was pretty big, but who cared about ships or planes or war stuff, what I needed was MONSTERS, their long dirty tongues getting caught up with the gear shifts and brake pedals of their humped-up wild kars. DADDY, The Swinging Suburbanite, mutated Tyrolean hat, blood-stained Glen plaid suit, green face. Commuting to Hell in a car made of a coffin, the hood ornament a wildly slopping martini. I don't know why Dad didn't hate this junk. He hated funny signs you buy at the hardware store, we don't swim in your toilet so please don't pee in our pool. Leering heart-disease humor that rose up behind our back yard from Jack

Hass's barbecue pool parties like smoky laughter at night, with the chlorine and tiki torch soot.

I got each monster as soon as it came out, as soon as I could save up for it. Fard made one and then gave up, I don't know what he played with, Noxzema jars. I had to have these models. They were a SERIES. You have to get EACH ONE of EVERYTHING. I liked having all the boxes, TO HAVE. I was lousy and sloppy at painting, I could only think about three weeks from today when I could buy the next model. You also had to keep buying the little jars of model paint, I could afford one of those a week. You had to have every color, and you had to buy new ones, FRESH ones of any color that might be running out. I only liked the jars when they were full. The way they display EVERYTHING of something in stores drives people crazy.

It's important to get stuff that's on tv. Honking around at recess we went through all the commercials to see what was what. Nunzio and Peter always had the most stuff as seen on tv, but since Nunzio STANK it didn't count as much as if Fard or Kurt got something. We all got jet fighter helmets, with birthdays and Christmas and Easter, some people get Easter presents. Jesus is still alive, here's Robot Commando for you. The helmets were all exactly the same, then Nunzio started saying his was better somehow, it came from a better store so it had *thicker plastic*. I hate Nunzio.

Arithmetic Town loomed over Saturday like a monster, it was grey and DOOM LADEN, but I couldn't wait to get my model home, lay out the newspaper and paints, make it in a sweat, get it out of the way as fast as possible. I couldn't stand it that it might lie around unmade, yet making it was going to give me no pleasure whatsoever and make me really mad. There was always too much glue, oozing out of the joins and all over your fingers, mixed with dirt and paint and then the tube of glue didn't look nice anymore and you had to worry about getting another one. Nothing looks like it does on the box.

Julie and I turned the corner, she was talking about RUBBER ANIMALS. I couldn't tell them apart, they all looked like melted erasers. I told her Captain Nemo ought to take them somewhere in the *Nautilus*. The low fences dividing the front yards one from another on our street are all the same height, even though they're made from different things, redwood rails, flower wire, ivy. Saturdays the dads of every yard walked back and forth, washed their cars, watered their lawns, talked to each other, fixed and painted their stuff. The way they talked to each other was weird. Neighbor Talk, Julie and I used to do it in the back seat of the car on a long trip. *See you've got something eating your geraniums. Heck, yes, got to get down to the nursery, thanks for pointing that out.* Ha ha ha ha ha ha ha!

Down at our house I could see Dad moving backwards and forwards. At first I thought he was mowing the lawn, but as we got closer the humid grey 'weather' pressed down and I started to think it was a kind of maybe NOT OK Saturday. He couldn't be mowing the lawn, he was pacing between the green car and the brown car, puffing on a cigar like a steam train. When we were almost there I got scared because I saw that the locomotive was huffing up and down the driveway with *Arithmetic Town* in its hand, held high like a flag. He held it SHOULDER HIGH so I would see it as soon as possible. Sheesh. What Mr Neighbor, raking right there thought, I couldn't guess.

Uh-oh, I said to Julie, I guess we shouldn't have gone to the variety store, now I'm going to get it.

Which was something we never said, *get it*, that's from cartoons or the Kellys. But I started to get mad, there was no RULE you had to do arithmetic first on Saturday. But there was the Permanent Rule, which I was stupidly ignoring, the rule nobody talks about, FUN LATER.

He looks mad, Julie. I didn't know I had to do arithmetic first.

Julie was always optimistic and encouraging, she wasn't afraid of Dad.

It'll be okay, she said, just say you'll do arithmetic now.

That's the kind of thinking these Julie people do, they don't understand how complicated and *full of crappy feelings* everything is.

I can't do it now, I said, it's almost lunch time.
So do it after lunch.

But he'll hate me all through lunch for not doing it, I said. If you can't go to the store or do a model before arithmetic, you sure can't have lunch before arithmetic, did you hear about that kid on Santa Ysabel who got real sick trying to do homework with *no food*?

No, gawked Julie, though maybe it was the idea of HOMEWORK that made her eyes bug out.

Now we were on the driveway. I tried to smile, tried to draw my model out of the bag to show him, distract him, anything . . .

Forgot something, didn't you, he said right away. He was wearing his safety glasses, no doubt he'd been hammering in the garage; in the cigar smoke his eyes beamed out like revolving Santa Fe locomotive headlights.

Yes, well, I, I couldn't make myself bring up the rules, what was the point, you try to say there is no rule and they shoot you down with one of the YOU KNOW THAT rules, you always finish school work before you buy wax lips YOU KNOW THAT.

Better get to work, he said.

Okay, I said, and saw that the day was so grey that the frontier lamp was on over the kitchen table, where I would shortly get beaten up by the police of *Arithmetic Town*. I'd be suffocating in one of those friendly family cars, Linda Johnson's family I bet, all of them with that *fixed smile* of hers, or mauled by one of the schoolyard dogs that smiled like Linda Johnson. Under the frontier-style light at the kitchen table, the harsh frontier light that made every duty sad, every meal angry.

I got rubber animals, Julie said to Dad in the outside world.

Dads want their sons to do what they have done. To be SWELL without effort. They want their sons to inherit the thing that makes *them* happiest. But they don't.

Under the frontier light, I thought my way up and down the street. It was true, Gomez was big but he wasn't going to be any Wrestling King, Larry was so dumb he couldn't even dress himself. The Kellys, said Mom, were *nothing short of criminals*.

Great-grandfather was stern by trade in a book-lined room, Grandfather loved football, Dad escaped football into arithmetic, I try to vanish from it into the cornucopia of the dark.

Some things work out OK, I thought, not great, but OK, even though this was a cloudy NOT OK Saturday, I could do the model later and that would FIGHT GLOOM. But the whole family was waiting to eat lunch at the kitchen table where I was doing arithmetic, where the horror light beat down on me every second. That started to scare me because Julie got cranky when she didn't eat and might turn against me. And Dad would have to use his large teeth to tear into his Monterey Jack cheese sandwich after stomping around in his gardening boots. I was getting some problems done but I started to study the nubs of the colonial tablecloth, they were shaped like the problems, I noticed. Dad came in, covered in impatient gardening dirt, then Mom shot in and started rattling pots she had no intention of using, if I have something wrong, I thought, he'll tell her to GET OUT.

I got the multiplication all wrong, I was guessing at the look of the numbers, trying to think how, then *trying to get out of my mind how*, the problems looked like the nubs.

Think you'd better have another stab at these, he chuckled.

Yes sure, now we can laugh, I thought, because LUNCH is on the way.

Why don't you do them this afternoon, he said.

Mr Generous, pushing my model and my little time, my little square of happiness, that's how I thought of it, a warm wood-colored box with a light in it, WAY to the end of the day, maybe into the night which was all wrong for making a model, it didn't have the right light coming through the windows.

After lunch I sat in my room, sickened by the OVERHEAD LIGHT which Dad put on when he abandoned me, just to make sure I'd probably throw up. Man, you know you're doing arithmetic, or having an operation, if that light is on. I looked down at *Arithmetic Town*, kept turning back and forth between the problems and the cover. The cover *was* the problem, that was EXACTLY the kind of town where Linda Johnson and all the people who can do arithmetic live, I bet that was how Linda Johnson actually thought of our school, and our town, even though we had no season like that, whatever the season on the cover was, or TREES. I started to get real mad and I also had to go to the bathroom, but if I opened the door Dad would be on me like a German shepherd, *Are you finished?*, I could just hear it. Say, he'd think, let's just go into the kid's room and see if he's having any fun by mistake.

I always have my table. That's what Fard says too, *neat table*. Everything I've thought about or dreamed up or tried to make, except for my machines, I've done at my table. It is an old table from the garage, painted black, though some of the wood shows through. It has a drawer underneath which smells like pencil shavings and ink. When I'm happy I sit at my table and I don't have to do anything, I share my happiness with my table and it shares its happiness with me. When I'm sad I *always* sit at it. I can open the drawer and breathe the pencil shavings and ink and I feel better. The way the black paint is worn off the corners and the knob of the drawer.

Fard feels better, he told me, by looking at the brass knob in the middle of his ceiling light. I know exactly how he feels but no OVERHEAD LIGHT would make me feel any better. I draw monsters at my table, I was sitting at my table when I drew the bad pictures with Gomez, that I completely forgot about in my drawing pad, the wiener and the bottom. In brown crayon with stuff coming out. When I have my sprayer I stand at my table, at my window, and look out, my window is now the window of Mission Control, I'm looking out at Cape Canaveral, not Mr Postum's yard with his blind dog Ralph yapping and falling off the kitchen steps. I'm glad my window is wide, I can really survey things from here, standing at my table holding the sprayer to my head. Nunzio has the same kind of window, sometimes that's Mission Control, Nunzio and Fard and I look out at Nunzio's pool, which Nunzio insists is *more like Cape Canaveral*. Because of the ocean. Nunzio stinks. When they gave us our school picture last year I colored his teeth in with black crayon and then put a big *X* over his head and wrote with the ball-point pen that has the school's name on it DO NOT LOOK AT NUNZIO'S PIKTURE, I spelled it that way to *add stupidity to Nunzio*. I've written my word, my secret angry word on pieces of paper at my table. Sometimes I hold the pencil in my fist and push down really hard to write the word, so that it almost tears the paper, that's the best way to write it, though if you push too hard you'll dent the word into the black paint of the table, which wouldn't be good, my word that is only ever said in private or torn up, you don't want your secret word of misery anywhere to be found.

Just to show myself I was master of my own room, I very quietly took the model out of its bag and

67

opened it up. It was a monster in an airplane, which was supposed to be held up by balloons he'd tied to the wings. There they were, balloons modeled in grey plastic. I thought I'd just take a little break and paint the balloons, then get right back to *Arithmetic Town*. I glued the two halves of each balloon together, really bad, the glue oozed from the joins and got all over my hands, mixed with dirt and paint, wait'll Dad saw. I'm such a crumby model maker that I started putting the paint on right away, maybe it was the wrong kind, Kurt would know, but true to the weather and the overhead light and everything about that Saturday the plastic just *melted*, went soft and the balloons that were supposed to look buoyant and funny, like on the box, looked like grapes somebody'd stepped on at school. Everybody hates fruit. So there was nothing to look forward to, my monster was going to melt, it was going to be a problem Dad and I would solve under *harsh light*, everything was becoming arithmetic.

I really had to go to the bathroom, but I couldn't go to the real bathroom. If I opened the door Dad would smell the glue. I went to my closet, where I had a round cardboard box. It would be neat to have your own toilet in your closet, I thought, it would be like the TRAIN where everything you don't need folds away and is hidden. I could tell Nunzio I had my own bathroom in my closet. I shut the closet door and in the dark I aimed for the round box, man, was that thrilling, wait till I could tell

Fard, who cared that it didn't flush. It would just go away like most things do. I felt good, and mad, at the same time. I went out to my table, grabbed *Arithmetic Town*, my wiener was still sticking out, went back in the closet and holding *Arithmetic Town* over my secret toilet, with steamrollers of sofa feeling going up and down me, I PEED ALL OVER IT. I wish.

The bell. The *bell*, I hadn't heard a word She said!

It isn't a cafeteria, although they call it that, it's just tables outside and this ugly lady with a big spoon. Really mean. You get a tray and she BLOPS your lunch on it. On hamburger day you take your triangle of cheese and poke holes in it with your straw and throw the triangle at Linda Johnson's hair. Then you can shoot people in the eye with pieces of cheese, from the straw.

Last month Fard got tuna surprise and there was nothing in it. Surprise, *Fard*, said Nunzio.

Fard and I had our still-new lunch boxes. I had baloney with warm mayonnaise, two cookies, the banana that is always brown. Fard, peanut butter and a cold BROASTED potato that sweated. Plus our stupid milk.

Peter sat across from me, eating the lunch for which he always got teased into the ground, deeper, into the mantle, magma, cold wieners without buns. His mom *lined them up* in his lunch box on lettuce, with olives and radishes. His lunch was garnished. Could you garnish my lunch, I once asked Mom. She went and got Dad.

Wieners made Peter lonely. His big house up on the hill was lonely, I thought, I went to see him up there and I could never figure out what was where in their house. They didn't move through it enough, no family events made dents in its atmosphere. In my room I wasn't allowed to wage war, but I could pilot the *Nautilus* or have Mission Control. According to the walls and the curtains I guess I could rope dogies if I wanted to. Cripe. Anita's room was full of Science and horse passion, Fard's of laundry. But Peter's room looked like the Boy's Room in the Sears catalogue, all the furniture matched, it was *units*. He had a *globe*. Peter didn't have stuff TO HAVE, he had things on display he didn't play with. He was an apprehensive guy, afraid to make a mess. Ironed jeans. Garnished lunch.

Peter had a teenage sister, who like Anita's brother was a remote figure. I'd never seen her but we often gaped at her bedroom from the doorway. Giant pillows and soft toys were perfectly arranged on her frilly bed. The pink world of dating. They had a LIBRARY, a high room where you could look out on one of the dry gullies. Grandma gave me a copy of *Born Free* for Christmas, I didn't like the story, it was about animals, and I didn't like the look of Joy Adamson, she looked like she was going to bite you. But I liked the book, mild charcoal cloth stamped with squares and triangles, I thought it would look good in Peter's library, I took it up there once to try it out. Even though I'd taken good care of it, when I put it on the shelf with their other books it looked tired and little.

Peter's mom acted like she didn't get the whole scene, Peter and his sister were just THERE. They had a Mexican lady who made the lunches, the whole place was clinical and CANOPIC, all the emotions of their family stuffed into jars in the library like the house was their tomb.

They invited me to go to the Rose Parade. I wasn't too keen, I preferred to watch everything on tv. *Why don't you go*, said Dad, *it'll be an adventure. You might even meet Bill Welch*. My favorite announcer. The idea of seeing Bill Welch's HEAD, which was shaped exactly like the top of the conical microphone tying him to his station. He was a microphone with glasses. White shirt. I got Dad to attach a piece of pipe to a really long piece of wire so I could be Bill Welch anywhere in the yard or even on the sidewalk. Getting up before dawn, Dad's idea of adventure. Although . . . Bill Welch . . .

I'd never got up that early except to stop my nose from bleeding or to throw up. It seemed 'cold'. It was really dawning, which seemed like something that ought not to be seen, when Peter's mom drove us up to the hospital. His father came out in his operating gown, which had red things on it. *Big crash on the freeway, no parade for me today. Bye boys*. Goodbye, I'd been scared that

71

this guy, the MASTER CARVER, was going to come with us.

Hours before the parade people were filling up grandstands and thronging the sidewalks, people with doughnuts and thermoses and those sheepy parade smiles. The taste of hot chocolate, which is not a taste but your tongue blistering and being burnt off, every half-hour Peter's mom tossed some more down us. You could pee in a rest room made of canvas and watch it steam. The 'cold' again.

The parade was confusing, it wasn't a SHOW like on tv, but a series of *boring insults* without glamour or light or sound, fat realtors on suffering horses (I thought of Anita, her indignation), stupid tributes to the Governor, the City of Cucamonga Says Buenos Dias America. Peter and I, as the only ASTRONAUTS in the crowd, were looking forward to the space float, but it made us SICK, a crescent moon made of lilies, crude planets of *stupid regular garden flowers*, a waving idiot in a suit from *Sea Hunt*, not Project Mercury. Peter's mom was annoyed we didn't like it. *Why I thought you boys loved space men.*

As the parade wound away, the floats getting even less inspirational, the bands tuneless, the twirling sloppy, she gave us lunch. Eat the cold wiener. They dropped me off at night with crusty eyes and a felt flag.

Favorite Teacher was on yard duty. In the hot

breeze, Her hair and skirt. The girls from our room were playing tetherball. Linda Johnson told them what to do.

Look, said Fard, Linda Johnson let Nancy Hoffman play tetherball.

Nancy Hoffman was playing tetherball like she plays everything, shutting her eyes and lunging forward with her arms straight out and her palms raised. She played kickball that way.

She looks like something out of *Hansel and Gretel*, said Fard. And look at Denise trying to hide her head. They play tetherball without looking and get the same score we do when we look.

Denise was one of the girls who wanted to be a teenager. She owned a black skirt with a poodle on it and a pink sweater and wore them every day. You could tell she had an older sister. In September she leaned over to talk to me during arithmetic. *You're cool*, she said. I was amazed, not only was she telling me right in the middle of class that she liked me, for which Fard would envy me, Gomez taunt me and Nunzio pound me, but she used the word *cool*, which had something to do with teenagers and the pink stuff. What, I said. *You're cool*, she said, looking at me in an odd *even* way. Now I began to choke, to joke, and I said No, I'm not cool, I think I'm pretty hot, and fanned my face Phe-e-w! Her steadiness became icy. *I said, you're CRUEL*. I felt like I was suddenly dropping things although I was holding nothing. Me? *YOU'RE CRUEL*, she said again, louder, what the, maybe she thought I was aloof to her beauty. *Because of that baby bird*, she said. I'd told her for some reason how Larry had decapitated a lost bird, I just found myself telling her. What I had thought was interesting about it was where did this bird come from, THERE ARE NO TREES, it's surprising what you find on the ground. I saw Larry in the gutter one day, looking down at this undone nest, and a tiny bird of skin flopping on the cement with its eyes shut. Everyone's always telling you that the HUMANE THING TO DO is PUT ANIMALS OUT OF THEIR MISERY, but Larry wants to put them all out of their misery right now.

Benedictine O'Hara wasn't playing tetherball, she thought she WAS a teenager. Everyone calls her B.O. She's almost as tall as Favorite Teacher, I thought. B.O. has this friend, Marilyn, in the other room, the other tall girl, they look at themselves all the time in their compacts, they talk about boys you never heard of. B.O. walks around like a lot of tall people, as if she's trying to pull her shoulders down to the ground, smash her face on the floor, pull herself under the earth. It's a very weird look and you can see it every day. B.O. seemed so pulled down when you looked at her it's as though the tops of her eyes had been planed off flat. She sat at a lunch table with her hands flat against the fronts of her knees. B.O. thinks that's ladylike.

The other girls, Rhondda, Roxie. Roxie was in my first grade class and I liked her name so much I started using it as my own name, at home, or

I tried. *Joe, come and get your hamburger,* Dad said from the barbecue, and Mom had to say, *This isn't Joe, it's Roxie.* Dad said, *Is it hell.* But out by the Big Tree, in the field, with the horses, was Anita.

She whinnied and shook her ponytail. Her hair is blonde, streaked with chestnut, she often wears it in a thick, perfect plait. I spend a lot of time looking over at it during arithmetic and also social studies, at the points where the hanks lap and flow one under another. In her eyes there is a luminous grey intelligence, *Science.* She's grownup in blue and green tartan dresses. Her black velvet shoes are often dusty at three o'clock from BEING A HORSE. The next morning they look like new.

Anita understands how you stick two plumber's friends on top of a cardboard box and it's a nuclear reactor. Mrs Dentyne went into Her closet of secrets. No one has ever been in a teacher's closet. No one has seen flash cards unbent, neat stacks of disks and slugs, towers of one-inch graphed manila. But you know these things exist don't you. Her turquoise lambswool sweater hanging neatly on a hook, boxes of new Pink Pearls. And She brought out to us the greatest junk, it wasn't junk it was STUFF, drinking straws, rectangular cardboard bands from bunches of pencils, horse chestnuts, doorbells, pie plates, broken tile, dead Christmas tree bulbs, halved tennis balls, pencils red at one end and blue at the other, circular

75

typewriter erasers, brads, rubber bands of every color, little rubber tires. And She said, *Today we're going to make Something out of Nothing*. I laughed inside, I had it made, this is all I ever did with Anita. Julie too. It's what I did in the garage when Dad wasn't around.

Dad was baffled by my machines. *What's this supposed to be?* It's a machine that—see, these are the controls. *What an unconscionable waste of nails*.

I took burnt-out fireworks from the alley and lined them up by height on top of a box, that was Captain Nemo's pipe organ. Gomez and Larry knocked it over and wrecked it.

Staring into Her Something out of Nothing box my mind roved over the stuff, it drew to me the things I needed exactly, it always had to be touching every kind of thing in the world at once. I took out three cardboard bands, a few straws, and two red tail-light lenses. I stood up the three (so important) cardboard things and put two drinking straws, crossed, in each. Then I put the two lenses on the table in front of them, where Eric had drawn military insignia in pencil and then smeared them with his own spit, our official solvent. *Obviously*, NUCLEAR REACTOR. Gathering, discharging power through its Tesla coil-type antennae. In a mighty chamber.

Fard had laid bent straws on his desk in an oval and drawn a chain of things on binder paper. *What's this?* She asked. A railroad. She came to me.

*What's this you've made here? Boys and girls, can
any of you guess what Joe's made here? No? What
is it, Joe?* A NUCLEAR REACTOR. *Hmm, well I must
say I don't quite see that.*

Anita saw, and she knew what ought really to be
done with the junk I ordered from the EDMUND
SCIENTIFIC COMPANY OF BARRINGTON, NEW JERSEY,
diffraction gratings, equal-volume solids, mirrorized
balloons. I could make them into anything, part of
my unconscionable waste, but Anita knew what they
were *for*.

She smelled of horses (where could it come from
but will?—she didn't own one), furniture, staid-
ness, England. I could find her in old wood. She
only smelled more that way when she exerted
herself, she was a strong runner because of her
long legs and because she WAS A HORSE. The smell
was in her house too. They lived up in the Hills,
which did NOT mean, Dad said, that they had more
money, the Hills were inconvenient and made of
mud. But they're high up, I said, defending her.
Anita's house was dim inside, full of things that
loomed:

The piano: Anita made me sit beside her so
she could show me she could reach more than
an octave. She played nine bars of Mendelssohn,
stopped and showed me the stretch again. While
we sat at the piano her mother *corrected our speech*,
from the kitchen.

The cars: a collection of valuable metal toys

77

from EUROPE you were not supposed to play with, except slowly, on the carpet. They might have belonged to her brother, a remote, aristocratic figure.

The shells: the garage and a room off it were full of giant seashells of horrible shapes and colors. This, Anita said, was her father's career, his specialty. His SPECIMENS. He was always to be addressed as Doctor. The shells were malevolent in sea-green cabinets, with peg-board backs and cut-out lettering. Even though you could never see the animals that lived in them, the pale yellows of the interiors of the shells made you as sick as if you had. Anita said they would be her career some day, she was already beyond discussing bivalves with me in detail. Her whole family was collections that could not be touched. When I thought about Anita and her parents I pictured them separated by miles of oak table.

Her room was lousy with horses, alabaster horses holding up real biology and astronomy books, silver stallions flanking her microscope, a big plastic horse from a carnival rampant over pinned butterflies. My filly, who knew what HYDROLYSIS meant.

The backyard was steep, which charmed me, from the cement part of town. She had swings and a slide as long as the one at school. From the top you could see across several gullies and cañons, houses perched at the ends of drives which were black paint on mud. You could see Linda Johnson's house, with a white colonnade which suited her with her pigtails so tight she couldn't close her mouth and she could write cursive in second grade already and never stopped saying it. We laughed at Linda Johnson because with all this talent she said she wanted to be a TEACHER, ha ha ha! it makes me laugh still. She only said that because it was awful, something you couldn't really be, like ASTRONAUT, SCIENTIST, HORSE.

You could see Kurt's house and the gully where we played on Saturday mornings if I stayed over with him. We would stay up late by flashlight suffocating to comedy records with pictures of night-clubs on the covers. Then early in the morning we put on our BROWN ZIPPERED JACKETS and stole in quiet through sofa feeling and tule fog for the gully. The gully looked like you shaped it yourself out of clay, like a Project, on a board, a few twigs or parsley for vegetation. At least you got mud, roots, gravel, a spider, something, but I felt more out-of-doors reading the works of HERBERT S. ZIM. On our block it wasn't even possible to get muddy. Kurt used the gully to torment his little brother, we hid from his cries, pressing ourselves into slimy hollows like dead saints. Crawling in the gully I got sofa feeling and I worried about Kurt's sisters, like his dad did, an intense disciplinarian shave-head. *We do not lock bathroom doors in this house!* he once screamed at me in greeting.

It was Anita's birthday and we were in her parents'

car. Orangethorpe, forever Orangethorpe, stores punctuated by citrus groves. I bounced up and down and talked too much for her father, I could tell he and her mother had marked me for VULGAR-IAN. They believed I had buck teeth and a crew cut, but I didn't. I had a *Regular Boy's*. The stores came and went, bigger and smaller, The Broadway, pets, a broaster, not Fard's. I made a commotion when we passed a store called ANITA SHOPS. You should buy all your clothes there, I said. They all acted perplexed and put out, like I was talking garbage, although Anita looked excited that a vulgarian was talking garbage in the Invertebrate Family car. Her father didn't snort, like Dad, but he breathed out for ten seconds. Her mother didn't like that I was mentioning a chain store, where of course people from the Hills would never shop. Anita's mother and my Mom both struggled against vul-garity, I thought. Good luck, girls. Anyway they all acted like they'd never noticed ANITA SHOPS before. But it had a big neon sign.

At the Mexican restaurant next to the variety store, where Mom took us when Dad went on business trips with his hat and overcoat, I got to stare close up at the neon sign in the window —the color doesn't fill the tube, the gas glows small, away from the glass.

The dusty parking lot. I wondered how generous Dr Mollusk was going to be, with that PhD of his. It didn't seem much like a party, just them and me. It was a 'cloudy' evening, the smell of tiny 'rain' drops in fine dust. Mom was glad we were having the party here, at Mr Knott's, she'd started yelling at Anita's mother on the phone about Walt Disney, *O, Mrs Mollusk, he is a Crook, he is a King, blah blah blah*.

I started looking around for Iner. I thought about Iner every bedtime, I made sure he was on my mind dial. He wore a derby and a big moustache and drove electrified San Francisco cable cars around the perimeter. Of all the machines of the place, this most of all. The grip levers, one red, the shiny brake handle. The clerestory, rich layers of paint over curved matchboarding. Being lifted by Iner to reach out and take hold of the thick rope of the gong, Iner lifting me into the heaven of the mechanism. One Christmas they gave me a plastic derby and a black paper moustache. I'm Iner, I told anybody, who figured I'd flipped again, thanks a lot.

The carousel they dragged here from the midwest, along with the 'Carnival of Venice' and 'Over the Waves'. A wooden knob on a rod beat the side drum. A bell rang when it was going to start. I always looked up at the tops of the poles, the greased cranks from which the beasts hung, I searched for the heavy grey feet of the carousel, under the slatted floor, and speculated on its big red motor. I sought the rich old electric smell, the best thing about Mr Knott, there in the popcorn and euca- lyptus. Anita got on the carousel with me but I remained lonely in my love of cranks and motor.

She seemed bored with the carousel, I didn't know if I should point out its real excitements. She tried for the brass rings, which had never occurred to me.

In Ghost Town we panned for flecks of gold in a sluice. An impatient geezer dressed as an old sourdough helped us. *Not like that, you little jerks*. What a job. But, I thought, Anita knows a lot more about gold than you ever will BUDDY. He gave us the flecks in tiny bottles and she started telling him about gold, metallurgically, it made his head swim and he called for her folks to come and get us off him. I knew the properties of gold thanks to the works of Herbert S. Zim. Her father wasn't interested, he *wouldn't listen* to mineralogy. *The ocean holds our future, my dear*. That's Captain Nemo's idea, anyway, Dr Mollusk.

I loved the Mine so much my flesh crawled, with big jolts of sofa feeling. I claimed for my own anything that moved on tracks, even car washes, the inevitability. In we went, artificial pitch dark, inside the cement and metal mountain. Gradually we made out distant galleries, miners moving mechanically in torchlight. The train snaked and creaked across a high trestle over a steaming pool, the water turbid and glowing within, you'd never be found. Pots of bubbling Yellowstone mud, a black-lit chamber of stalactites and stalagmites— there in the dark only Anita and I, and Herbert S. Zim, knew the difference. Our teacher didn't. Do you? I loved the Mine because you couldn't see how it worked, never a glimpse of wire, floodlight, the cement floor beneath the track, hoses, speakers. The cave-in at the end, terrified men screaming, the beams above you bending and splitting, I turned around to watch everything returning to the way it was, the ceiling going back up, rocks ceasing to wobble, ready for the next train. I loved the Mine because I knew its secrets.

Anita's parents wouldn't go on any rides. Her father's shellfish retraction, her mother's flight from vulgarity. To get to the restaurant we had to walk through Ghost Town and under the Leg, from an upper story a lady's leg protruded, hinged at the knee. Every so often it jerked.

I don't understand that leg, said Dr Mollusk.

Looking at the leg always started me worrying, it was hard to credit as entertainment.

Don't you know what it's for? said Anita, flicking her braid in a way I liked, it meant something interested her, even if it was just how stupid you were. It means in the wild west you're supposed to go up and kiss them.

Who, said her father, fretting.

Dancing girls, Daddy, she said. Probably.

Her parents were in a daze, this, their child's birthday, was just something to endure.

Two Americas warred in Mr Knott's dining room, adobe walls hung with bulbous Indian objects vs. red and white checked tablecloths, baskets of biscuits, the grim overhead-light dinners of the pioneers. Dr Mollusk ground his teeth through the biscuits and crispy chicken of this obligation, these terrible things that dragged him away from his own personal inky blackness. We were God's own irritant, grit in his blue-point flesh. Boysenberry pie, he didn't even relish that, it was merely a grim reminder that he was away from SHELLS.

Why do you do that with your pie, he said, we do not eat pie like that.

Anita didn't care for food, a sterility, as from her house, hovered around her tartan lunch box. Whereas Peter's lunches were garnished and mine had always got too hot.

In pie torpor we meandered toward the car. In a window a guy was making huge suckers out of coiled candy rope which came steadily out of a machine. He wound it into disks and stuck them on pointed sticks while wearing a paper hat. Dr Mollusk was palpably afraid we would want these, no, afraid we would want that guy's job. We walked by the livery stable and Anita paused in the warm breath of it and smiled.

Orangethorpe, even longer now, in the 'rain'—I bet you I remember every day it was 'wet'—the ANITA SHOPS sign was off. They delivered me home and her parents stared at our house. I could imagine her father saying to her mother on the way home, Why do people live in *houses?* I thought to myself Anita hadn't enjoyed herself, she hadn't NEIGHED. Thank you very much for the very nice time.

*   *   *

84

There was always a moment when we knew our minutes were numbered, up in the Hills, an hour till dinner time and they were going to come and get me. Up there was something you didn't feel in our neighborhood, a BREEZE. Anita never played at my house. Her mother.

Their swings looked like no one ever used them. Anita's brother was remote and aristocratic, but not athletic. Anita was slender from BEING A HORSE. I never saw her on the crossbars at school, she had Nurse Wrists anyway, as Fard pointed out. The knowing hour descended with its chill, we looked at each other and got quiet. Anita climbed up the slide. I wondered if her swings were firmly rooted in the sand, it was hard to imagine Dr Mollusk doing any construction work.

Come up, she said, in the authoritative tones of the Scientist Horse. Come up. You slide under me and tell me what you see.

I began to choke. This wasn't right, not with her, not the star of my show, my movie. Of course, Fard's sister had shown us, in the resinous trailer now one with puckered swimsuits.

Slide, she said.

Slide. I got on my back at the top of the slide. Craning my neck forward I looked out at the colonnade of Linda Johnson, Kurt's gully, his sisters, Anita's dress blowing above me. I slid, far from mighty or sofa feelings.

What did you see, she said.

Kurt's house.

What, get back up here, she whinnied. Didn't you see anything you weren't supposed to see?

No.

She was discombobulated, she couldn't understand that I wouldn't look. Real doubt crossed her face, had she disappeared under her dress entirely?

At her mother's call from the house we wondered if her tone was disapproving, if she had been watching us. Our car circled in the driveway and our mothers exchanged cool, surmising pleasantries. Anita was still mulling it over—what the hell had happened down there?

Out beyond the horses, Gomez and Larry were in a corner of the cyclone fence. I have to talk to Gomez, I said to Fard and Nunzio.

Fard said, did you know Larry is so dumb he only comes to school at recess, for Gomez to beat up. They worked out a deal with the Principal.

Gomez was doing something to Larry with a banana. He jabbed at Larry's face, which was wary with vaguely remembered injustices, a LAB ANIMAL'S face.

Larry is under the influence of evil spirits, said Gomez. Aren't you, fucker.

He stuck the end of the banana into Larry's mouth. Larry must have been afraid of something at Gomez's house after school, out of earshot of Gomez's mom.

Gomez, I said.

*　　*　　*

86

On Saturday I was following Dad around, I had seen it was a Technicolor day when I got up and I knew he would garden, what he called gardening. He put on his pith helmet, shorts and gardening boots, lit his pipe and whomped up a big pail of dichlorodiphenyltrichloroethane, as he liked to call it. I was preventing him from thinking, probably, and talking a lot. What are you going to do this morning?

Wellsir, I'm going to water the geraniums, and then I'm going to get after that pesky crab grass, and then I'm going to go to the nursery and buy some more geraniums, and then I think I'll have a Monterey Jack cheese sandwich, what do you think about that.

Jesus Christ, I said.

He was walking ahead of me and he stopped, his pipe fell into the bucket of dichlorodiphenyltrichloroethane, *phut!* He grew and grew, he was going to get me, last week I'd called Julie a *shet-ass*, like Gomez was always saying, and he'd gone crazy, I didn't know, obviously, what that meant either.

Do you even *know* what shit is? he'd said.

No.

It's what comes out of PEOPLE'S BOTTOMS when they go to the bathroom.

What!

And listen, Jesus Christ is a name for God. God is terribly important to some people and they don't like it when you shout his name around.

87

Is that why you don't go to church with us, I said, you don't want to shout the name of God.

No. I don't go because I don't, because the lawn, *forget why I don't go*. We're talking about you, not me. I mean, we're not talking about that. I'm talking about these bad words you're using and I'll tell you one thing, Mister, I'm going to get to the bottom of this right now.

I followed him through the house, tramping, banging, looking for Mom, GOD was involved and it was terrible.

I want to know where the kid is getting this language, you know what he called his own sister and now today he said Jesus Christ right by the fence, for Pete's sake. I heard Jack Hass drop his pool net and laugh at us.

Boo hoo hoo.

Are you happy now, you have made your mother cry.

Boo hoo hoo, it's that awful Gomez.

Gomez was bad, wait, I agree . . . Gomez's mouth scared me. His lips were full and small and you could usually see the tip of his tongue. He was usually smiling because he'd said something dirty. Even in class Gomez said dirty things under his breath. In art they gave him clay and he made a wiener, in alphabet soup he spelled BUTT and PIPI. His elastic shorts were sweaty.

I saw a sixth grader bully Gomez because he was fat and wore zoris. *Hey, kid, that's what dirty*

Emexicans wear, that's what the stupid people in our town call them, *E*mexicans. That was the only time Gomez cried, he broke into a run and I was surprised it wasn't to get his dad. He just ran, away. His dad was a gritted and lubed mechanic and wrestler, GOMEZ GRANDE. His growling crouch and flabby smile appeared on posters every month at the Santa Fe station.

Gomez's mom had his face, sort of. She used to be the lady that walked around the ring with a sign between wrestling matches, puncturing the canvas with her shoes, blinking at hoots. What she did now was SWEAR. When Gomez Grande came home from the garage he usually found their house full of laundry, why was there so much laundry in our town, and he swore a lot at Gomez's mom, *Caramba, bitch*, and she swore right back at him, *God bless America you big shet-ass* (that's how they said it and that's what I called Julie, a *shet*-ass, which upset Dad even more, Why are you *pronouncing* it like that?) *you think I haven't been working my fingers to the, Jesus Christ*—So Gomez really had his mother's mouth on him. He had a sister, too, darker, she had a thin face. Gomez's mom swore at her for her darkness and kept her in the back of the house, always in flouncy dresses and white shoes, like a communion doll in a box. Their house smelled of starch, nail polish, hair spray and cigarettes. None of Gomez's toys had batteries. Larry was his only working toy.

*　　*　　*

89

Gomez, that little fat kid, said Dad.

He's—

Don't contradict me.

You ought to see the drawings, said Mom.

What drawings, Dad and I said together.

You bring your father your drawing pad. No games. Be quick.

EVIDENCE: Gomez had been over and there in my big drawing pad were a bottom poohing and a wiener drippling pee in burnt sienna, on a dare. Gomez did this, I said.

I'll bet, said Dad. This is very bad. I'm going to go over there and give that guy a piece of my mind.

Who, Mom and I said.

Why Mr Gomez, of course. Mr Bad Words.

Mom got all pale. He's, ah.

What?

I don't know. A wrestler.

Dad stopped in the doorway, looking at the wire telephone-stand in thought. Wrestling . . . Come here. Listen. People who use bad words are often violent. So when you go to school on Monday I want you to be careful and tell Gomez that these words are very bad and that I disapprove of them very strongly. He shouldn't be saying them and I don't want you to associate with him anymore. And if that's a problem for him, it's just *too bad*.

Then a big heaviness came over our house, the way they looked to each other, their happy marriage rocked by a VULGARIAN.

Sometimes you see into adult land, Fard pointed

out, when they quit their act, when you see them
stare at their furniture and fish bowls and cars as
though they're the junk that they are.

Larry whined and let the banana leave a big track
all over his face.

The evil spirits told this little shet-ass to suck
on this thing whenever I say, said Gomez.

Gomez, I said. My Dad got really mad at me,
I called Julie a shet-ass, do you know what that
means, and I can't play with you anymore, so
goodbye.

Gomez's eyes got real bright, that mouth darker,
wet.

You little shet-ass. You know what's going to
happen to you? The evil spirits are going to send
someone to get your father because he's a shet-ass,
and your mother, because she's a shet-ass.

What evil spirits.

We got evil spirits, don't we, Larry, said Gomez,
shoving Larry's face.

Yuh there are evil spirits, Gomez.

There are not.

You meet us after school at the Big House, said
Gomez.

You stupid id—, Nunzio goes.

The bell.

She was smiling.

I've managed to find a television we can watch.

Yay.

91

We'll have to share it with Mrs Sagerhammar's class.

Yay.

So anyone who wants to watch the space mission, please clear your desks and get ready to go. If you do not want to watch the space mission, you may stay here with me and read quietly or draw.

What if you don't want to watch or draw.

Well, Julius—

She's not coming, whispered Fard.

What do you care, you fruit, said Gomez.

Boys boys.

We went to the Big Room, the floor was gritty.

I hate to sit on the floor, I said to Fard. I have to keep changing between kneeling and sticking my legs straight out.

How come you can't sit like this, said Gomez.

There on tv were the Science Reporters, busy, important, white shirts with their sleeves rolled up. Diagrams and dark clocks against a flat map of the world. The Name of the Sponsor.

Man, said Fard.

This isn't CBS, said Nunzio. Teacher, can we watch CBS, it's better, he said, looking around at everybody.

This tv doesn't get channel two, said Mrs Sagerhammar, in her brown suit as usual. Let's watch this, shall we.

Guy, said Nunzio, how can it not get channel two.

Now, what's your name.

Maybe it's the antenna, said Simon, pointing at a wire flopping over the edge of the tv, wrapped in Band-Aids.

The A-V guy strikes again, said Fard.

Now we're going to Mission Control, for their hourly report there, and we understand that they may speak live with the capsule on the air. We take you now to Mission Control.

—is Mission Control. Time is 0945, hour five of the mission. *God Bless America* is now passing over the continent of Africa—

I sneered at Linda Johnson.

—fifty miles above the surface of the earth, at a point approximately above the city of Durban.

Durban, said one of the white shirt guys, if we can put that on our—

God, said Nunzio, on CBS they have an *automatic* map.

Young man—

Mission Control has just informed us, as you have heard, that the Colonel is directly over the city of Durban, in Africa.

This is Mercury Mission Control.

Mission Control coming back on now—

In a few moments we will be speaking with the capsule live. Please stand by.

Well, Mission Control has just informed us that we will be able to hear them speak with the Colonel, it's been a number of hours since they did this, the relay station in Fiji—

93

This is Mission Control. We will now switch to the Flight Director, who will make contact with the Colonel in *God Bless America*.

Man, said Fard.

Mercury, this is Mission Control. This is Mission Control. Do you read.

Jer. Fission cun.

So much static, here they got the guy up there and you couldn't understand him.

Ector iner ive, plete ickin ory of sih ittbar enema.

Man, said Fard, what's he saying.

Julius put his hands over his enormous ears and leaned toward the tv up on its stand. Try this.

Roger, Mercury, this is Mission Control. Colonel, we have a special greeting here from the President. On behalf of all Americans and the people of the world, he wishes you well and Godspeed.

Aww. Ih seber nor kibble up to ere. Oud mercan izah ih ah ewcut. Eehive wife. Fuc.

Huh?

He says he's proud to be an American up there, said Julius, covering his ears completely, no small task. Try it this way, it really works you guys.

Then those of us in the know, those with SPRAY-ERS, covered our ears when the Colonel spoke from space, thanks to Julius's discovery. They played a film loop of the Colonel's HEAD moving back and forth a little, as if it alone was looking for something to do.

I couldn't move anymore, my legs were already numb, so I could not escape the pee trickling across the floor from Julius. Nunzio and Fard moved out of the way without a thought. Julius had given up even trying to work on his timing. This was too important to leave, I saw that. But. The raw feeling Julius was going to get at the edges of his shorts, walking home. We heard the Colonel clearly, saying hello to his wife, barking to his dog, complaining about his back.

Must be time for him to pee, said Gomez, they gonna tell us how he does that?

Who said that, said Mrs Sagerhammar.

Suddenly I was lifted up into the air, roughly lifted and shaken by the arms I was using to hold my hands over my ears, dumped down and turned abruptly. The enormous red face of the PRINCIPAL.

What's the idea!

What's what idea?

You know what I'm talking about, my friend, said the Principal. His bow tie palpitated like it had a squeeze bulb. The buttons on his white shirt interrogated you like headlights.

Your teacher let you out of class and all students came here especially, to the Big Room, to watch the space mission. This is history, son. You must be goofing off. I think we'd better have a talk.

He expected me to follow him to the door but my legs were ANESTHETIZED and I slipped in Julius's puddle.

Say, quit your fooling now. Wow, who did this.

Julius started crying.

You'd better go to the office, son. The nurse.

But, said Julius, the mission—

Never mind outer space now. You think the Colonel wants you to wet yourself? And you, what's your name.

Joe. Lake.

Joe, what is going on here? I want you to come with me.

Ontrol, o eelings in my eggs, said the Colonel.

The Principal dragged me through a waist-high ocean of crappy feelings into his office, which he tried to keep the sun out of with slitty blinds. There was big-guy copper stuff on his desk.

That kid Julius wet this boy in here, he yelled at the secretary, get us a towel. I think you're being very rude, he said to me, very rude to your teacher, who let you come down to the Big Room on this special day because something historic is happening. I think you're being rude to fellow students by horsing around when they're trying to learn about space, the newest frontier. To *learn*, Joe. And the worst thing, you're being rude to the President of the United States, who's paying for all this with my money, and your folks' money. And you're being rude to the very brave Colonel, Colonel—

Colonel Bleep, I said.

—the most important and special American there is today. What if he came to visit us at our school, to tell us about outer space, and you sat in the front row with your hands over your ears. WHY DON'T YOU WANT TO HEAR THIS HERO?

I had to sit in the Office for an hour, *thinking* about it, watching the secretary, Mrs Bitchman, jangle her jewelry which was like a part of her, her perfume and teeth and gum. I sat under the MASTER CLOCK, its row of buttons like an accordion, this is where they make the BELL ring, I thought. Julius's mom came in with a new pair of shorts, a pile of shorts she wanted to leave with Mrs Bitchman. Mrs Bitchman said she didn't have room anywhere in the Office but Julius's mom and I said *You'll need 'em* at the same time and she reached out and grabbed at them. Julius came in to change his shorts. Two second graders came in.

What do you want, said Bitchman.

He bit me.

Honestly, is this how this year is going to go, why did you bite him. Biting isn't for boys. Puppies bite. Boys don't bite.

I could hear the Principal in his office: But as a parent I think you should look at the other side of this, ma'am, that growing up in a world crisis can be character building. Why just now I had a boy in here who *closed his ears to space*, he didn't want to *know*.

This Principal is a DOPE, I thought, of course I wanted to know. If only he knew when the distributor cap in the brown car cracked, Dad gave it to me, and I put it on my table and slowly brought my eye down to it, doing my crane shot. It had

97

battlements, its curved surface was broken by a flat place, a portcullis in my scale.

O this was definitely going to the tub.

My knee rose a Gibraltar out of the soapy water. I got a mighty feeling as my eye traveled up to my knee. There the distributor cap sat, majestic, I'd forced it onto my knee cap, as a castle grips a hill, as the mighty Griffith Observatory grasped the crown of Los Angeles. The heat lamp was on in the ceiling and if I plopped at the surface of the water with the flat of my hand, reflections of the red element shot around as if they were flashes of lightning in the sea. Dad came in.

Hello son. Nice bath?

Dad, what are these anyway.

Ah. Someday those will help you—hey, what are you playing with that gunky distributor cap in the tub for?

It looks like the Planetarium.

The what!

Can we go there tomorrow?

Dad didn't like astronomers, he thought they were escapists and fakirs. He complained about them a lot at dinner, and after reading the paper. *I tell you, dear, it's not responsible. If they're not gazing off into space, they're just sitting at a table and counting, counting, counting. . .* Then he would go to the sink and fume there, you could tell he was fuming by the way he suddenly washed his own plate, banging it against the sides of the sink.

*   *   *

98

The two of us in the early kitchen. Juice and cereal. *I don't think we'll stop for snacks today*, said Dad. The Planetarium was sacred to me and I couldn't associate it with food anyway. We usually got a hot dog in the park, far from the marble Science halls where I would hate to dribble anything and would DIE rather than throw up.

Los Angeles rose off the plain, you had the feeling you were *entering a city* as of old. Going up the hill in the park was like driving up Mount Palomar, both with DOMES at the top and Dad driving beside me. HERE WE ARE, Dad's incantation. He took my hand and led me up the walk, up the steps and into the vast central hall of the Planetarium, my favorite place in the known world, *Principal Dummkopf*. Everything there fed my dreams and nightmares.

In a black granite well the Foucault pendulum traveled back and forth, so heavy and sedulous that I didn't think of it as swinging. Beneath it a clockish dial, Roman numerals (I learned to read them there) and pegs the great brass ball knocked over every quarter-hour. As some people are drawn to and then over the lips of office buildings and the Grand Canyon, my mighty urge of sofa feeling was to get in there with that ball. It looked like Death, a thing moving back and forth in a tomb. Would you die if you went in there, I asked Dad. *Wellsir, I don't know about that, but you would get thrown out of the Planetarium FOREVER*.

Down one of the echoing halls we watched people

outside through the Camera Obscura. They were upside down but still slurping ice-cream and itching their crotches. Precisely because they didn't know you were watching, I thought, they acted doubly stupid. Their behavior embarrassed me in front of Dad. But here in my favorite place in the world was what I feared most: down another hall, past illuminated black and white photographs of galaxies spinning and black-lit portraits of the giant planets, was a darkened room, with something in it you might at first take for a person—a cone of wire coiled six feet high topped with a metal ball with two spikes sticking out of it. The first time I saw this, the giant Tesla coil, I knew I had been running from it my whole life at night.

When the guide came to demonstrate it, violet lightning shot out of the spikes and crawled the walls of that horrible room. The Tesla coil hummingly illuminated neon tubes he held in his hand, though this was no more frightening than when you stared in at it, alone, when no one was near, coiled there powerful in purple gloom. What would happen if you went in there, would anyone be able to go in there ever again, or did they have to seal the Tesla coil into that room forever, hoping it was happy to sit there and not *reach its evil blue hands snapping voltaically out and around?*

Wellsir, WE'RE ALL SET. Showtime.

Phew—!

The hush of the matte black doors, then we saw the silhouette of the city and the yellow sky.

100

Music, a scherzo, skittered around the dome and through the girdered armature of the ZEISS STAR PROJECTOR, my favorite thing in the world in my favorite place in the world. Some kid at school called the Zeiss projector 'dumbbell-shaped' and I wanted to sock him in the stomach. You're dumbbell-shaped, I said. So great was my love for it that I watched *it*, in the dark, and not the night it made, watched its steel spheres rise and turn, their many lenses glowing, and the opaque projectors in its church-furniture base dim on and off with comets and rotating nebulae. I watched specially for the white mouth that projected the moon in its phases.

Calmly like a doctor the Voice of Science took his seat at the controls, which I got to look at, if only Fard had been there, an acre of glowing dials turning, each a recreated star, larger green indicators, POLAR ALTITUDE, SATURN PROJECTOR, ECLIPTIC, an electric schema of the Milky Way. The Voice of Science washed over me and Dad like a prophet's, like a cure. We felt significant, tiny, wise, fools, powerful, dead, worth-while, trumped-up, hopeful and clownish each in their turn. That is the only place I've ever seen the stars.

Welcome to Griffith Planetarium today, ladies and gentlemen. Take a moment to look around you. From up here you can see all over the City of Angels, and you can see a lot farther than you can outside. You probably noticed there's a lot of that pesky Southern California smog out there today.

101

Pesky, said a lady in back of us, why, don't they know that it practically *killed* Betty.

Get on with it, said Dad. Dad had lain on his back in the north woods and seen more stars than anyone in Los Angeles, I thought.

Our great Los Angeles basin used to be a fine place for getting to know the stars and constellations, but these days it's a little harder. So using the magic of our Zeiss projector, let's roll the hours forward to around dusk tonight.

Ooo!

You see, our planetarium is a sort of time machine. With it we can see the skies of next week or thousands of years ago. Now, as darkness falls, let's watch the moon rise and the first evening star appear over Los Angeles.

The dome above us turned orange, then bloody red, violet, then dark blue. The silhouettes of the Planetarium and the TREES in the park disappeared. The projector whirred and stretched within its cranelike cage, the moon projector poked its glowing smile out of the Canopic-looking torch where it lived and the moon rose, to a scratchy hi-fi nocturne. A bright point appeared in the west.

Ooo!

This is Mercury, the planet closest to our Sun, and the evening star in the western hemisphere this month.

There in the black sieve house of the dead I was enveloped in the Voice of Science, listening to it intently but not to what it was *saying*, Dad got mad at me on the way home, *What do you mean you didn't hear him explain how to find Arcturus?* HERBERT S. ZIM and my beaker and Project Mercury, Anita and Tutankhamun all swam with me in the air-conditioned night. The smooth see-sawing of the projector, the stars which ran across my shirt front before they died on the floor is what I watched, not the stuff happening on the ceiling, which might as well have been millions of miles away as it was for true. I wanted to live in the Voice of Science, it was Canopic, it was mighty, in a Milky Way like that I would give people lots of sofa feeling. I wanted to light my room so it looked like a dome.

Build me a dome, Dad, I whispered.

Huh!

That's what I was thinking about when cracks of light appeared under the exits and I knew the ushers had opened the outer doors of the light-locks, how could I have this at home, this soft dome light, this holy machine with its struts and mysterious lamps *in my room?*

We went out into the sunlight made vague and hard to bear by the yellow haze all over the city. People wandered around in the forecourt, eating ice-cream, scratching themselves and not knowing, or not remembering, they were being watched by folks inside the Camera Obscura. A sallow man in a dirty white shirt, a battered hat and brown pants appeared from behind shrubs. He had bags

103

under his eyes and hadn't shaved and he smelled like Julius, like pee.

Hey buddy, he said to Dad, you speak English.

I stopped in my tracks to hear Dad addressed in this way, people only said *buddy* in the movies. Dad was surprised too and turned. No, he said.

Gimme a cigarette.

No, said Dad, the way he said it when he meant it. He got red in the face and reached for my arm, not embarrassed to say no, embarrassed to be asked.

He means it, I called back to the smelly man, who looked at me and went back into the shrubs.

Imagine that, Dad said to himself, brooding in the brown car, a bum. A *bum* here in Los Angeles.

Several days later Dad began cutting pentagons out of cardboard. When he had twenty or so he painted them black. Then, with electrical tape, he put them together to form a ball. He made a base for this and put a lamp inside it, I was going crazy, huge feelings lapped at the back of my neck, cold as a tidepool you fall in as the sun goes down at the beach and everyone is already waiting in the car. He took out our star map, a wild blue and yellow thing which he had declared would Never Be Properly Folded Again, and started sticking the black hedron with a thick needle from the sewing machine. We went into my room that evening. Dad pulled down the window shade and drew my cowboy curtains. He plugged in the black thing and I looked up at the ceiling. There were Ursa Major,

Orion, even the Serpent Holder—he'd poked lots of them into the thing.

And which one's this, and which one's that, he was asking me.

I don't know . . .

Weren't you paying attention at the Planetarium? What do you spend all your time with this map for.

For the colors of it, for the Voice of Science, I couldn't tell him that. I could hardly wait for him to get out of my room, I was looking at my ceiling as though it were the dome of the Planetarium.

We can learn these this way, said Dad. It's not a toy.

Mom made noise in the kitchen.

Well, have fun with this now, he said, and went out.

I put *Gaîté Parisienne* on the phonograph. Welcome to the Planetarium today, ladies and gentlemen, I said. Take a moment to look around you.

Yeah, *take* a moment, I thought, the second graders were in with the Principal and he kept saying the same thing Mrs Bitchman said to them out here, puppies bite, boys don't bite, puppies bite, boys don't bite.

The bell. I guess you can go back to class, said Bitchman.

They were all there, looking at me, Julius and Nunzio, who'd had their hands on their ears too. No transistors, no more Project Mercury. A crackdown.

And having been made to sit in the Office during recess, which shows what they really think of you, now—

Let's get ready for PE class, She says.

Marching out to the field, Kurt and Fard and I walked with our arms out in front, our mouths mechanical, shooting our eyes quickly back and forth. We did this around Favorite Teacher because She looked like the beautiful space blonde tv puppet Dr Venus. We might have to RESCUE Her at any time. *You know*, Dad said, *if you boys put half the study into your arithmetic you do into that idiotic puppet walk*.

Now we had to play a game, not one you'd ever heard of, but one that came of course out of this thick maroon PE book all teachers have which I HATED to see on their desks or coming out of their bags. Divide class in half and line students up at 2½-foot intervals facing each other, eight to ten yards apart. A thick moron PE book from a college where they specialize in PE, according to Fard, the theory of kickball, that had to stop, I thought. We were all in the clothes we would watch rot on us through the year, acquiring paint and holes and worn edges and scuffs, becoming so much laundry. Maybe Fard's mom would do a lot of it. But this was the first time She had dragged us all the way out there in the dust, it felt like we were going to play this game in birthday party clothes, Linda Johnson and Roxie and Anita in extra nice dresses, I was confused as usual.

106

This exciting game was one side throwing soft-balls at the other. *Then, the reverse*, that was the touch of brilliance that made this game worthy of inclusion in *The Thick Maroon's Treasury of PE*. Always you stare into the sun, the PE sextant guarantees that. Avoiding it I looked down into the dust and crazy stems of crab grass that made our field an uneven kickball hell. To see the others you had to hold your hand over your eyes *and* squint. We started out throwing, Nancy Hoffman was across from me so I would have balls to chase but nothing to fear. On it went down the line, back and forth, She trilled Her whistle if you threw to the wrong person or missed the ball.

You mean there are points in this game, said Fard.

Oh yes, Fard.

It was funny but Fard had the idea, or the boredom, to bring his mitt to school. It's hard to catch a softball in a baseball glove, doubly hard for Fard because his was a catcher's mitt, puffy, orange, *folded*—the first time I saw it I thought it looked like a member of his family.

Suddenly Fard's dad was going to buy him a glove. I walked slowly with him and Fard around a bin of mitts at the variety store. Fard's dad would pick one up and pound his fist into it, like he could increase its value, its look. He walked around and around the bin and looked up while he pounded them, grimaced and gazed into the upper corner

107

of the store, or heaven, for—money? He squinted at me.

What kinda glove you got, Joe.

Rawlings.

Rawlins? That a good one?

Yeah. I didn't know, it never softened up, I thought, even though I wore it, punched it, sat on it while I watched tv. I spit on it, oiled it with special stuff. Dad bought me a softball to go with it so maybe part of the problem was that my mitt had never caught anything, just grasping stubby pain when the ball pranged the ends of your fingers.

Once I had a mitt I got the idea Dad and I should play catch. *I thought this might happen*, he said. I guess he had been hoping I wanted the mitt just TO HAVE, which would have made him cranky too, I could just imagine, *You get something you USE it damn it in the way the Rawlings Company intended.* The neighbors were surprised to see the two of us come out after dinner, with a BALL. Neither of us knew where to stand. Jack Hass stopped cleaning his pool and peered over the fence to watch us. He was a guy with chest hair the same color as his skin. People started coming out of their houses to see me with a mitt on my hand. *Hey, everybody—the Lakes are playing catch!* The thin leather strips along the fingers of the mitt made me feel funny.

Dad overhand, I underhand.

Ha ha ha, said Jack Hass.

We are very bad at this, I said.

Dad took a stab at the vocabulary. Bingo. Hot diggity.

That is not baseball talk, I said.

How do you know? he said.

We were so bored with it already, we couldn't hide it from each other or from the clutch of neighbors, Mr Postum, Mr and Mrs Neighbor, Jack Hass, *Larry's* dad, jeez.

Pretty thin stuff, huh, said Dad. Let's call her quits.

No, no, called Jack Hass, throw higher, you can do it.

Don't give up, Dr Lake, said Mr Neighbor. Come on, Joe.

I felt like Herbert S. Zim playing catch, here I was weird again, thanks a lot. Dad paused and looked back at our neighbors. *Sorry*, he waved to them. *Sorry*.

This new kid moved in across the street and introduced himself to me by asking if I wanted to play catch. I pointed at Jim and Jimmy's house and started to back away from this kid. *The guys who live THERE like to play catch*, I said. Cold but honest. My mitt was only a grim reminder.

A dark shape eclipsed the sun in one of my eyes, like the shape that lunges at you when you're trying to go to sleep, I heard the whine of a shell in wartime cartoons, the big hard softball thrown by Nancy Hoffman with her eyes shut that was knocking me out, the guy standing there thinking how lousy he is at baseball, here he is in the Office, *You again?*, for the second time in one day. But now one of the Glorious Dead, truly dizzy with ice pack and having had a glimpse of OSIRIS, She leaning over him on the doctor-type bench, concerned and annoyed and anxious to make the teacherly point that you have to pay attention or look what happens.

That was sort of a wild pitch, She laughed, like a kid. Her forehead glistened. I think you'd better lie here until we go back to the room.

She had a word with Mrs Bitchman. I lay on the doctor-type bench and watched swinging flashes of light come under the door when anyone entered the Office. The Principal kept telling Mrs Bitchman to do something and she kept telling him she didn't have time to do it. The door opened and closed and the phone rang all the time. Then Murry McCollum the ice-cream guy filled the Office with his high-pitched fat voice. Every day he hunched in his MICROBUS with his head shaved, in a dirty white shirt with a tiny bow tie, a cold invincible White Owl in his mouth.

What are you people, trying to put me out of business.

Now Murry, listen, we told you you would probably need a permit this year, said Bitchman.

I have permits coming out of my ass.

Whoa.

Would you watch your language, Murry? This is a school.

110

What *permit*.

A vendor's permit, from, the city said you would have to pay.

No. They told me you just apply.

But then we, the District, we have to charge you—

You people can't make up your minds, you are driving me crazy. Don't have charging if there was no charging. If you are going to charge, charge. If not, don't.

Well I'm sure I don't, the Principal is—

Why did I just get a ticket. Why is there a big cop with his big foot on my bumper. And what the hell is this Trailer of Prayer that's right in my usual spot. How come *they* don't get a ticket.

It's the PTA. Murry. They just—don't like you selling that kind of thing there. At that time. At this time.

That kind of thing, it's *ice-cream*.

You know, it's, they're concerned about the kids' teeth.

You ought to be worrying about your teeth, sister.

Murry! There's no call to threaten me. They want to sell juice bars.

Juice bars? They're already dumping a ton of fluoride in the water. They want to put me out of business so they can sell their own ice-cream. Then everything will be hunky dory in Costa del Khrushchev. You tell those little Hitlers I'm going to sue them for restraint of trade, see if I don't.

There was silence in his wake, maybe Kleenex being pulled out of a box, then typing. I'd never heard Murry say anything other than *Thanks, kid, careful crossina street*, his gesture toward the big time, Good Humor.

They're always snarling about teeth, I thought on the doctor-type bench. The best thing was to go into the bathroom and run the water down the drain for a minute and wet your toothbrush. Then come out smacking your lips. Everyone's happy. Kurt said he rubbed his brush on the edge of the sink for the sound of scrubbing. He had to, his Dad was always listening at doors, for sounds of creeping maturity.

Twice a year Mom took us to Dr Dennis Boer, a dentist for children, I called him the DENTAL BORER. He made Julie cry. She cried whenever she heard CHILDREN'S menu, CHILDREN'S shoe store. *Highlights for Children* was in the waiting room, and *that* made her cry, its creepy ideas and terrifying pictures, comic strips of hideous wooden puppets and on every page a lesson being worked up you: Goofus Rudely Refuses to Help the Pitiful Old Lady, But Gallant is Polite: *No, madam, I'm most dreadfully sorry, but I shan't be helping you and your unpleasant rash across the street*. The Borer's paddle-ended fingers in your mouth, his buck teeth, soft crew cut, his WOODPECKER'S FACE, his *act*: Now, hyuk hyuk, I'm just gonna reach back here and see what I can see, hyuk hyuk! Maybe I'll find buried treasure! Do you think so? I'd have to be a old cuckoo bird to think that! That's me, nutty as a cuckoo bird! Hyuk hyuk!

When he leaned hard on your jaw the drill slowed and you could smell your own teeth coming to the boil, Julie always went first and you could hear her non-stop screaming. The Borer's stupid clowning took a new turn when he got down to the PULP.

Oops, guess I found a nasty old cavity back here! Guess I'll have to patch up that stupid old cavity! Won't take a minute, hyuk hyuk!

Then he leaned forward.

Now Joe, I want you to know there's something you can do to lessen the chances I'll find cavities in your teeth again. There are some bad people in the world, Joe, right here in America. They want to put things in the water we drink. Chemicals, Joe. They say they will make your teeth stronger. No one knows anything about them. They might change the way you think, or look. You might wake up in the morning and BE AGAINST Mother and Father and Sister and Teacher and Reverend. So I want you to tell Mother, Joe, that you think it's bad to put anything in our clean California water. Will you do that for me.

Well, what are you supposed to say. He's got a drill. But I thought to myself, I'm in favor, all in favor, of Chemicals, and Science. And so is Fard. You creep.

Okay, sport! Guess what! I've got that cuckoo old cavity fixed up now good as new, hyuk hyuk!

Trying to figure out how many times he was

113

going to say *cuckoo bird* was the only thing to look forward to. Then the Prize Shelf, if you didn't bite or claw him you went home with a ring-toss game, a misprinted doll, a red and yellow horseshoe magnet, or a sorrowful-looking little man printed on a card, an inch of fine chain for his nose. Tapping the picture gave him lots of deformities to be sorrowful about.

All right Mother, said the Borer, we'll be sending you our bill. We thank you much, and here's some literature for you, Mother.

The Council on Dental Therapeutics of the John Birch Society, *Fluoridation: Eroding Your Teeth's Freedoms*. After we went to the Borer the last time Mom mentioned him to Dad in the timid way she brought up Gomez Grande.

I'm not sure it's a good thing either, said Dad. Once you set a precedent.

Fard said, I'm here to take Joe back to class.

Joe Lake? He's in the nurse's room.

Did you see Murry a minute ago, said Fard on the way back, because he looked really mad.

Murry's in trouble, Fard. But my Dad is on his side, at least I think so. Or else the dentist is.

What dentist.

Either the dentist or my Dad must be on Murry's side, I said, they couldn't be on Bitchman's side.

But what's wrong with him?

I think it's zoning.

114

Fard? Joe? You boys need to take your places quietly while we wait for the bell, She said.

Standing behind our desks, Fard and I looked at each other, we were always a little sorry to leave Her. Without rescuing Her. She likes to make us be quiet for the minute before the bell rings. The red sweep, the flag. The handwriting chart. I heard Julius's plugged nose.

The bell, and in the rush She said, how are you feeling now, Joe, do you want me to call your mother, so she can give you a ride?

No.

All right. You look after him, Fard.

I will.

Sunlight.

That was close, I said, we have to go see Gomez at the Big House.

What for, said Fard.

He says he has evil spirits, didn't you hear that crap.

What about the meeting?

We have time. We can just stop off and see them.

Are they small.

I guess. Hey! Where's Murry?

Quick, said Fard, look behind the Trailer of Prayer.

But Murry was not there. A lot of people were standing around in shock. Now it seemed hot. Hot as the yellow of the crossing guard's cracked cap. In turning forty-five degrees, east, home, in the wind, a little te deum down my shirt.

115

Uh-oh, said Fard.

Hey, *Tutankhamun* has nosebleeds too ya know,
I said. My Dad took me to see his junk in a museum
in Los Angeles. They had the jars they put his
GUTS in.

Man, said Fard.

He had gold tubes to put on his toes, with the
toenails molded right in and EVERYTHING, I said.

That day I said to myself, now I'm going to see
the most beautiful STUFF there is. If someone will
lift me up. I never had so much sofa feeling, looking
at Tutankhamun's chair, his stone headrest. Then
I found his alabaster head, the King. It was a lid
from one of his Canopic jars, pearl white and soft
and alive. It had been painted once and the edges
of his nostrils were a dark crusty red. Buddy, I
thought. Of course he had nosebleeds, I said to
Fard, the dry air of Egypt and after you die they
stuff a lot of salt up your nose, after they *pull your
brain out through it with a HOOK.*

Man.

So you probably got nosebleeds in the afterlife
too.

Why are you getting all these bloody noses all
the time, said Fard.

It's the dry air, I said. Also I'm worried.

What about.

Gomez. Murry.

I worried about the taste of slimy blood going
down the back of my throat, and about waking
up in the night with my face in my pillow soaked

116

with blood. But once I could get in the bathroom and talk to myself in the mirror I was all right. I knew exactly what to do. They showed a film in Science last year, it was an Encyclopaedia Britannica Film. EB at the beginning, everybody falls asleep. I stayed awake for this one by sticking scissors into my leg under the desk, starting last year they gave us pointy scissors, because it was going to show about blood, and, I thought, I often have a lot of blood all over me and my stuff. Our doctor, who hates me, told me to stick ice on my nose but last year I used a hot wash rag because the EB Film *proved conclusively* by dribbling blood into big beakers of hot water and ice water that the hot water clotted it faster, but you know who the clots are, the people who make EB Films. I spent the WHOLE YEAR standing in the bathroom with a cramp in my neck, holding a hot wash rag to my nose which pumped frantically like a fire truck when I put the hot rag on it, so that I thought my nosebleeds were getting worse, that all the blood in my head was going to glug lower and lower and suddenly I'd look in the mirror and I'd have a colorless, transparent head, a VISIBLE HEAD like on the Science shelf in Mrs Sagerhammar's room, and the worst thing about that is the circular muscles around the eyes. All this because I stuck the scissors in my leg and watched that EB Film, those shet-asses with their extra large beakers. *But hot water does clot blood faster*, said Anita, *I happen to own the Encyclopaedia Britannica*. I'd

117

stand in the bathroom for hours, taking the wash rag off to make it hot again under the tap and to see if blood was still flooding out of me. Finally it would stop, I don't know who stopped it, HORUS. Better than OSIRIS. But then I had to get rid of the blood clot if I didn't want my nose to solidify into METAMORPHIC ROCK, I'd had that experience, thanks a lot, walking around with a huge flaking boulder in my head and making everybody sick when I smiled, big brown things falling out. But what was I smiling about.

Almost every day in March and November I got a nosebleed on the way home, it was arithmetic and tests. Fard would always come in my house with me and stick around till it was over. Because he's my best friend. But after I stupidly shifted to the hot wash rag, I'd move back and forth on my feet in front of the mirror, real slowly bringing my head back to the vertical, and announce *I'm going to blow out the clot*. Fard started to weave on his feet and he said something like *Blawaowaww, think I'd better go*. Who could blame him. Some of my clots wouldn't go down the drain and I had to poke them with the end of my toothbrush, which was another thing I didn't like. I ought to have had a set of Egyptian embalmer's nose hooks and a chest of Canopic jars to put all my clots in. Mom took me to see this doctor who CAUTERIZED my nose so it wouldn't bleed so much. I'd never heard of him but he had lots of jars in his office, which could have been Canopic. *Why do you think you're*

118

getting all these nosebleeds? he said. Arithmetic. Then he jammed this ball of silver nitrate up my nose with a stick.

Aren't nitrates what Captain Nemo felt he had to destroy because they were so evil, I said to Fard.

Think so. Yeah.

It felt like he'd wedged the planet Jupiter up there. *You'll have to leave that in for five minutes.* If I ever see that guy again, I said to Fard, I'm driving a railroad spike into his forehead. Who cares if he's a doctor? I'm back to ice now.

That's a relief, said Fard. There they are.

The Big House and its palm tree loomed over us. But you could never see the Big House or the palm tree unless you turned and stared right AT them, it's hard to see things that are old. They're the only thing in our whole neighborhood older than we are. Someone in a TOMB refused to sell this dark green shingled mansion. Its grounds were huge heaps of dirt clods, mixed with remains of an ornamental drive, broken up as if bulldozers came to the very door of the house before the skeletal hand was raised, just like the crossing guard's, in its mausoleum, to defy eternally redevelopment of the orange groves, its fortune, *wa ha haa*. We said it was haunted but that was only our duty because it was old and empty.

Gomez stood in the middle of the grounds, Larry peered at us from behind the palm tree. I watched the effect of the cement chunks and dirt clods on

119

my school shoes as we stumbled up to Gomez, he was brimming.

You ready for the evil spirits. Hey, he yelled at Larry, get over here.

Larry approached slowly, really afraid of something. There couldn't be evil spirits, I thought, but Gomez wouldn't have made us go to the Big House unless there was something. It might be a trick, I said to Fard.

Yeah except, said Fard, you know the ending of Gomez's tricks, you have to show him your bottom, or he socks you, and there are moms in cars right over there, on the street.

These guys want to see the spirits, Larry, said Gomez.

Larry slowly lifted a chunk of cement with cut-off iron rods sticking out of it. He bent over and started to dig, here and there, with his hands.

God you have a fat bottom.

Here they are Gomez, said Larry quickly. Gomez pushed him out of the way, WAY away, and pawed at the dirt, still dry six inches down. Our county. Small green things began to appear, vegetable, three or four inches long, with faces carved in them. Threatening, idiotic. Gomez looked at me, dark, wet.

This is them. They're angry.

Larry looked back and forth between us. Fard and I didn't know what to think. This was a really weird thing for Gomez and Larry to have done.

What do they say?

They say someone's going to come and pound your dad and then your mom, because they're both shet-asses.

I imagined it, opening the front door, someone's pounding Mom, what are you doing here, oh, oh, call the police son. Like who is going to pound them?

You don't believe me, shet-ass? They could make my dad do anything.

What would Dad say in a situation like this. *We'll just see about that.*

Gomez merely snorted. Larry tried to curl his lip but drooled on himself. Fard narrowed his eyes at Gomez and Larry as we backed away, he's good at that, it's part of his folded heritage, even though we had to go to Pupae with Gomez in an HOUR. We headed for our street. In silence. I know Fard felt indignation for me.

Here was a LEAF on the sidewalk, Hallowe'en was coming, even though you had to go to the variety store or look at the decorations at school to tell. The supermarket had a ghost in the window in wash-away paint. We had started loading up on wax lips, though. That's what made us best friends in the first place, the day in first grade Fard opened his lunch box and he had a baloney sandwich, milk in a thermos, and wax lips for dessert. Me too.

You get your costume yet, said Fard.

Yes, I've got my costume. Cemetery Pete. It's supposed to be some kind of ghost. It has an OK

121

mask but the body part is this *sack* of slippery grey cloth with a stupid picture on it that you can't see when you buy the costume in the box, NO TRYING ON COSTUMES, that's what the sign said at the variety store. Cripe, when the grownups open the door they're going to say *Oh, goodness, what have we here? Cemetery Pete!* I don't like that my costume SAYS something.

Yeah, said Fard, but that's better than when they say *What are YOU supposed to be?*

Fard usually had a sheet, laundry, with some junk drawn on it. He saved up this year and got a robot costume, the mask is good, it's square and looks a little like metal. The body part has nuts and bolts printed on it, it doesn't *say* ROBOT. Fard's dad wanted him to go around with just an apron from the Broaster, and hand out cards, but Fard stuck up for his rights. I think his mom saved him finally.

Man, said Fard, look, a leaf. It's fall.

The same way I made my movie about Anita, I caught corners, glimpses, angles of things where our town must be like other places. I searched hard for those moments and even tried to make them, I spent a lot of my time squinting at the eaves of a house in the late afternoon sun, or the TREE with orange leaves on the way home from school. I was looking for east coast elements, eaves and leaves, here, there. I put together a menu, my RECIPE for looking at things, things I look at in a WAY, an order, to get Hallowe'en.

There is a time of day that *is* a holiday. Hallowe'en is at five-thirty, when you look up and down the street, when the sky is dark blue, tending black. Most of the dads are home and people have lights in their kitchens. It's getting cool but no one is trick-or-treating yet, only one or two pumpkins are lit in a window or on a porch.

Thanksgiving is at three o'clock, when dinner is in the oven and the house is hot, and everyone gets COOKED into what they are, you look out the window and your dad and your uncle become the pleasant chore they are doing, the beer and sandwiches on the porch table, the quiet sunlight and their pipes, maybe you can hear the east coast on the radio.

Christmas is slippery. It might be at eight o'clock on Christmas Eve, when you're going to read a story and maybe you hear carolers, but the whole thing is drawing down, focusing on your bed. Last year I couldn't sleep and kept turning my light on and off, looking at my watch. I kept setting it ahead to six o'clock, just to see what six o'clock would *look* like when it came, when I could go get Julie. Julie always sleeps through Christmas Eve, she sleeps through everything, it's infuriating to see her face that doesn't move even when you start saying her name. Then I'd set my watch back to what I thought was the correct time, *allowing for* the time I imagined had passed in looking at the pretend six o'clock, consequently I made everybody get up at two-thirty in the morning. In the snapshots Dad looks like someone painted big circles on his face.

Christmas might happen at secret times on Christmas Day, not at two o'clock in the afternoon when people finally come out of their houses after they've been screaming at each other for seven hours. The grownups are getting REALLY MEAN around then. It might happen at breakfast, I caught a glimpse of it there once, between a piece of fruitcake and a tangerine on a green plate.

I have a WAY of getting the exact Christmas I want. In the garage there's a spray can of artificial snow I saved up and bought. They never let me use it, Dad said it's for STUPID PEOPLE. He forgot I've never seen any snow, so how dumb is it. Mom said you can never get it off the windows. I sprayed a little on the lantana the year I bought it and I thought it looked good, but Mrs Neighbor came out and said *That is not a plant for flocking, JOE*, and gave me that look. It's not even *hers*. I go and get the can and stare at it, because it has the best red and green. Looking at the brick-red and pine-green on the can of artificial snow you can think of everything about Christmas all at once. Christmas really happens the day Dad lays out the Christmas tree lights on the floor and turns them on, to check them. The red, green, blue, yellow and white bulbs, bright against the carpet. I don't need any of the other junk about Christmas, I don't even want him to put the lights on the tree, I want the lights on the floor for two weeks. Or they could string them on ME.

124

Man, Christmas STUNK last year, I said to Fard, I didn't get Robot Commando.

I know, Fard said.

He still felt bad for me. Robot Commando was supposed to be the thing we ALL got for Christmas, we made PLANS. Frank Armbruster was this kid in the other room who organized stuff, he was the guy who made us sit down and *decide* we were all going to get Robot Commando, so we could have a giant Robot Commando mobilization. I didn't really like that idea, I thought that would probably wreck my Robot Commando or at least make it dirty, let alone that I probably wouldn't get Robot Commando, but I got caught up in the Robot Commando fever and then messed up all Frank Armbruster's Robot Commando plans *by not getting Robot Commando*. I think Julius didn't get Robot Commando either, or else he got Robot Commando and immediately wrecked it so bad it was as if he didn't get Robot Commando. The worst thing was having to TELL everyone WHY I didn't get Robot Commando, which was that Mom got a load of the Robot Commando box at the variety store and said Robot Commando was *too violent*. Even though it was just sitting in its box. This was another way I was weird, thanks a lot, that I had no armory. I thought Robot Commando would impress Dad with its mechanical brilliance, it moved out to the middle of the floor, its eyes rolled up into its HEAD, the top opened up and rockets fired out at your enemies. Robot Commando's arms shot

big shells out too and it had treads, I yelled for
Dad to come in and watch Robot Commando on
tv and all he said was *Hm*. Then he walked out
of the room like Robot Commando had made no
impression on him.

My house looked about the same, the lantana, the
green door. Nothing had changed while we were
at school. I'm starting to think I don't want to go
to the meeting, I said to Fard.

Oh, it'll be okay.

But Gomez. And *Nunzio*. Nunzio's mom.

I know. But. See you down there.

Okay. Goodbye, Fard. Fard was right, how bad
could it be. I didn't like the idea of not seeing him
for half an hour, but I had to have a honey sandwich
and put on my uniform.

Mom and Dad picked Pupae for me, I wanted
to be in Wolflets or the Hiawathas. I didn't even
understand the word Pupae until Anita explained
it to me, *don't you remember*, she said, *we saw
regular pupae in EB Films*. Well of course I had
been asleep. Nunzio's mom said she was going
to found a Nest for our neighborhood. Mom and
Dad got more and more enthusiastic as we went
around getting the official junk, it was disturbing.
We went all the way out Orangethorpe to get the
Pupae Official Guide, which smelled like paint.
I started to CRY on the way home. *I'm stopping
the car*, Dad said. I was in the back seat reading
about the First Order of Merit, *Tree Climbing*. I

126

knew there wasn't a TREE big enough for anybody to climb in our neighborhood, so there went my first badge right there. *There's a tree at school, isn't there?* said Mom. Nunzio's mom will never go all the way over THERE, I said.

The more Pupae belongings they got me the more I started to think I couldn't possibly join, how can you let them decide everything? I started to say so once in a while, maybe Pupae wasn't so great, or maybe it was great but it was not for ME, but Mom gave me her *you're going* of brick when she tucked me in at night.

I took my time eating my honey sandwich.

You *are* going, you know, said Mom. She started sticking me into my uniform. Ow. You're going, she said. She made my bandanna very tight around my neck. Off you go, you're all set, Joe boy.

Okay. I don't want to go.

Nonsense.

I would show her who was master of my afternoons. I would Fall Off My Bike.

Goodbye, Mom said.

Wait, I said, ah—I was worried she wouldn't stay to see me fall off. You might have to call an ambulance or something.

What for? Goodbye.

Goodbye, I said and flipped the front wheel and now before her I lay in a dramatic tangle, the blue of my uniform, the red of my bike. I reached my hand out.

Are you all right? she said.

Like a dope I NOD, but reaching, reaching—this is PATHOS, don't you get it?

Better get going then. You'll be late.

And she shut the green front door! I said my secret word of anger on my bike. Now I had to go to the meeting with chain grease on my uniform, I expected Pupae Demerits. Fard and I had been reading about demerits. They were all a lot of us were going to get out of it. I was tempted, for a sec, to turn my bike toward the variety store. My Pupae dues would buy a whole bag of Army men. Wax lips beyond your wildest dreams. I passed Fard's house, his trailer, his bike was gone so he was already at Pupae. I heard the washer and dryer.

In Nunzio's driveway I pushed my bike over and jumped up and down on it. The garage door was open. Our Nest. There were a lot of Pupae, wriggling, looking around with their PARTIALLY EVOLVED EYES, according to Anita pupae experience *many anatomical changes*. Well Gomez looked the same.

Against the blue of his uniform, Julius's face was a red always astonished beacon. I made for it. His uniform was incomplete and untidy. Demerits coming. But he was holding an ear of Indian corn, which made me sweat, I didn't know we were supposed to bring anything except our uniforms and our dues. And a willingness, as it said in the *Pupae Official Guide*, to learn certain crafts, songs, and ways.

What is that Indian corn for, Julius.

128

If you asked Julius a question all you got for a minute was wet workings, I waited while he operated his mouth and his eyes watered.

This is all I could find.

Julius's life must really stink, all he could find. What are you talking about.

I couldn't find a banana, my mom doesn't have any and she wouldn't buy me one.

I looked around, no one else had Indian corn or bananas. Nunzio came in and headed right for Julius.

What's that, Julius, your *snack?* he said.

Julius rushed at Nunzio with the ear of Indian corn over his head.

It's not a snack, you crab, we're supposed to have a banana, it's part of the uniform STUPID!

Nunzio pushed Julius away. He looked at Julius and smiled, REALLY MEAN. Bandannas, *bandannas*, you're supposed to have a bandanna, not a banana. *Julius*.

Estupido, said Gomez, his eyes bright.

No wonder Julius is in the slow readers, said Fard.

Julius looked around at us and got real red. He went into the corner and made a snorkeling noise, I guessed he was crying. Julius is a real corner kind of guy.

Then Nunzio's mom came in. This was going to be tough, look how Nunzio was standing behind her, like a pole.

Having Nunzio's mom as our Nest Mother doomed

129

our nest, I thought. I was secretly glad but Fard was worried. We could see her toenails and they were red. Her hair boinged. She wore an official blue skirt, very tight, it was PIPED with Pupae white. Could you see Nunzio's mom throwing a line across a rocky gorge or taking us into caves, showing us how to whittle bandanna-slides in animal form as shown in the *Pupae Official Guide*? No. She had trouble perching on a high stool.

Hello boys. As ya know, we will begin with a pledge of the flag.

What flag, said Fard, looking around.

Since there ain't much of us, we don't have enough dues funds yet ta buy a flag. So for now we will salute this as our flag. Nunzio, bring the flag honey.

And Nunzio with his molar grin marched up with this bathroom fluffy pink rug! He held it up in front of him and then he nailed it to the wall. We were all embarrassed, even Gomez thought this was really weird, you could tell.

Now we will say the pledge, said Nest Mama.

Salute, salute, said Nunzio. Obey!

What are we supposed to say? I Pledge Allegiance to the Rug of Mr and Mrs Frank T. Lupo? But everyone droned out the pledge like at school. The rug was stained and had pieces of toilet paper in its tufts. Nunzio went into the corner and poked at Julius.

Come out of there, stupid, my mother commands it.

130

Julius, said Nest Mama, c'mere dear.

Nunzio pushed Julius to the central oil stain of the garage. The leaves of the Indian corn hung down.

Whatsa matter Julius honey?

Nothing.

Make him salute, Nunzio shouted. He barked in Julius's ear, Salute! Salute!

No it's all right honey, we hafta start the meeting now, said Nest Mama. Now boys, as ya know, the first badge in the hanbook is climbin a tree.

We rustled. We felt breezes, the NATURAL WORLD.

But I have spoke with the captains in Bellflower and cause there ain't no trees big enough ta climb here we are excused from this badge. Isn't that nice?

Man, she's *relieved*, I said to Fard, it isn't *fair*.

Are we going to get consolation tree badges, asked Fard with his hand up.

I thought we were goin to a forest, said Gomez under his breath, his mouth and eyes dark and dangerous and wet. In his disappointment he emitted POOR SMELLING GAS.

I've been out in the natural world, I thought, because of Science, because of Dad, because of HERBERT S. ZIM. If the sun shone warm and early on the soapy Natal plums outside the kitchen window, you knew you would wind up in Trabuco Cañon, the big smell of sage up your nose with a lot of dust. I decided to collect rocks when I dedicated myself to Science. All through the natural world,

131

nestled in the groins of riverside mountains, under oaks, TREES, in wide places in the road, were ROCK SHOPS. Inside, row upon row of snow-white boxes with cotton in them, each displaying a polished SPECIMEN unlike anything Herbert S. Zim could have known. Some of them weren't rocks at all but fossils they made out back, or colored glass. Usually the guy tried to make himself look like a prospector, dungarees, beard, hat, even if he got his rocks from the EDMUND SCIENTIFIC COMPANY OF BARRINGTON, NEW JERSEY. All along the highways I watched for rock shops and Dad knew I was watching. Right in the middle of nowhere you saw the sign, ROCKS GEMS MINERALS CURIOS POP and he knew he would have to stop.

Dad bought me a small cabinet with drawers in it for SPECIMENS, a word I learned from Herbert S. Zim, and a mineral hammer. Out of an old Desert waterbag Dad made me a holster for it. When I put on my holster I felt mighty and Scientific. When I thought in the tub about my hammer in its holster I got a mighty feeling. It was pretty small. I also got a mighty feeling if I thought about Mighty Mouse in his fortress. Or about the word mighty.

What do men together, hiking Scientific men, do before they reach their site? They make sandwiches at an early hour, when clear morning light shines dully on the kitchen tile, when mothers haven't stirred. Braunschweiger for Dad and honey for me. They stop at the variety store for Bonomo Turkish

132

Taffy, new banana flavor, as seen on Sheriff John's Lunch Brigade.

Oh all right, said Dad.

It was mysterious to be in the car, at the variety store early in the morning, the clerk drinking coffee and smacking her lips like sleepy people do in cartoons, as she added up the candy of men together.

I'd better have one of these, to keep you company, said Dad. WE'RE ALL SET—his cantillation.

Out on the highway. My view: sky and telephone wires, the dashboard. From our neighborhood of streetlamps, bushes, cement and dust to a world solely of dust, and blue sky above the dust. Out to the edge of our county, where the roads alternated between asphalt and gravel.

We were men together because we were driving in the *brown* car, a car Dad bought for driving to the Lab. Mom could have the *green* car for shopping. The brown car was like the earth and Dad's old clothes. The brown car had a stick shift, a secret pleasure of men together, and cloth upholstery which smelled of the past, and storms.

Out in the middle of the orange groves was a big factory, made of tubes, dim lights way inside it. What's that?

Wellsir, that's where they make the charcoal for our barbecues.

The factories of our town made fruit DRINK, not juice, potato chips, and trombones. Most dads worked in Santa Ana where they made submarine parts, fighter planes and ATOMIC BOMBS. Fard and I could see the fruit drink and potato chip factories from school, but you never got to see them load or unload potato chips or fruit drink, they were SECRET. *Everyone would be asking them for stuff all the time*, said Fard. But you could smell them, around quitting time you could smell the slide oil rising off the trombones waiting for their railroad car.

Charcoal, though, that was mysterious and here was where it came from. Since it was black and oily I thought it came from Ohio, a place I put together on my mind dial from the black engine, the tender, gondolas and other junk in my train set. I poked hot coals in the barbecue, the only open fire I ever saw, and looked into their orange hearts. Horrifying stuff went on in there.

Beyond the orange groves we came to a carrot farm, they processed the carrots right there by cooking and cooking them until you couldn't think straight. Then the hills rose up off the dry flat floor of our county. Humps with dots of scrub. Dad drove with confidence through the tiny cañons, the walls crumbly with shale, which didn't interest me. A guy who really wanted to hammer away, but there was no granite or limestone for him to hit. Probably a lot of geologists would be surprised to hear this, that the hills of California were made of DIRT. I would tell them.

Herbert S. Zim showed in watercolor how you collect SPECIMENS. You have a hat and glasses and a white shirt with the sleeves rolled up. It helped to have a hat and glasses. I didn't have a hat or glasses but I hoped I would need them both one day, for my career, like Anita's father's career, only not as MEAN. You have a mineral hammer and you get out there, in the field, and you chisel something off something else. You put it in your canvas satchel. You eat your Monterey Jack cheese sandwich out of brown paper. You wrap up the specimen in the sandwich paper and take it home and determine its place on MOH'S SCALE. You look it up in *Rocks and Minerals* by Herbert S. Zim. Then you paint a small yellow disk on one of its lesser planes and when that is dry you label it in India ink with the specimen number, and enter that in your serious-looking canvas logbook. I had a small binder I bought at the variety store, exciting and blank. I was shy about asking for the paint, Dad was afraid I would PERMANENTLY HARM THE GARAGE. He alternately opened and shut the windows of the brown car.

Sure can get dusty out here, son. Funny how all the scenery is on the edge of the county. Like a big broom swept it into a corner. He brooded while driving for a minute and then said: to ready the Southland for millions.

We went up and up the small cañons, wiggling in our car like sperms in EB Films. Only one ever makes it, but in the brown car we counted as one. Whatever they are. We were looking for the one place in Trabuco where there was a dribble of water so you might say you'd seen something.

A cattle guard and a gate and a black and white

striped reflector, the end of all the roads of the west.

HERE WE ARE—second incantation.

There were three smells, the overpowering sage, dust, which coated my shoes in a way I liked but then began to worry about, and as we hiked up the cañon, wet clay. Can you believe that out there, in the madroño and a lot of plants so dried up they didn't even have NAMES, a meadowlark sang?

My hammer got hot in the sun and it banged against my bare knee, singeing me, as I tried to keep up with Dad. I kept looking for tough rock to hammer, but only the trembling shales surrounded us, you could have used a feather duster, not a hammer. This was the only rock you could collect in our county, outside the white boxes lined with cotton. The heat and the heady smells and the rhythm of the hot hammer against my leg, no water, the colors. . . Whenever I was out with Dad I began to worry that maybe I was afraid of Science, that even if I were eventually awarded a white shirt and a hat I could not be Herbert S. Zim. Driving down Mount Palomar from the mighty telescope, Blugggh! In the depths of the Sea Lion Cave, Oregon, Blugggh! In an elevator at Pacific Ocean Park designed to make you think you're going beneath the sea, Blugggh!

Time for a break, I'd say, son.

Bonomo Turkish Taffy, grape soda, water from my metal canteen which was boiling hot. Dad began to brood, looking at his waxed paper bag.

I ought to treat myself better in the morning.

Was he remembering being alone in the north woods, when he was ten? Or didn't this dry place remind him of that at all? He must always be disappointed in our county's landscape, I thought. He said he wanted to *introduce me to country*, but this really refused to be countryside. You just felt swept aside, with the hills. Then we were off, Dad talking again, I'd asked him a question. *So when the combined weight of the molecules—* He stopped and looked at another gate, leading off the road. Instead of a reflector, a big NO TRESPASSING sign. Beyond it the walls of a side cañon led away from us, the curve of it romantic.

How about going in there, it looks promising.

But everything they told us at school, obey warning signs, don't talk to strangers, don't break the law, multiplicator, multiplicand welled up in me, my fear I was failing arithmetic, the hot grape liquid and Bonomo Turkish Taffy, new banana flavor.

Blugggh!

Hey!

He laid me down in the sage on the shaded side of the road. He sheltered me with his hat.

Why don't you ever eat any regular food? All you had was a lot of nauseating yellow candy, whenever we go somewhere you throw up. It's television. X-rays have destroyed your equilibrium.

We moved upwind of my puddle, now nearly evaporated on the hot gravel road. Miserable, I pictured Herbert S. Zim crawling along the road, his hat bent and forlorn, his glasses steamed up, his white shirt with the sleeves rolled up blotched with taffy spit and thick grape RESIDUE. His logbook dirty and his specimen bag dragging in the dust. Residue is a word Dad would use, I thought, like when he used the word MEMBRANE to a waitress, referring to the surface of his pea soup, and made her faint.

Kind of hot out here, isn't it.

Yes.

You didn't get any rocks with your hammer.

This is all shale. But maybe I could hammer the driveway at home, I thought, there are pebbles and pieces of shells in that.

Wellsir, what about a visit to the bird sanctuary?

My *yeah* still shaky and dehydrated but full of relief. The sanctuary was a deep cool cañon full of TREES and chirping birds you couldn't see, but at least they were important enough to have names. You could buy things there, packets of bird food, hummingbird syrup, from a friendly man with suspenders and his sleeves rolled up. It's scary, I thought, when you're away from things to buy, or from PRINTING.

It was hot in the brown car. I worried I'd thrown up on Dad's good time, but maybe he thought Trabuco was dull. You can't figure these people out. I liked coming home late in the day, to see the life of our county again, cars moving through the orange groves, past oil wells. To look proudly at the dust on my shoes and my hammer, even though I hadn't used it. To think back on the heat in the small damp cañon. To hear Dad say nothing to Mom about my throwing yellow and purple up. My hammer WAS our trips into the hills. But what made me happiest was to sit and look at it under the electric light at my table.

So-o, said Nest Mama quickly, as ya know the second badge is Arts an Crafs. An today we are gonna make a Arts an Crafs project.

We breathed out toothpaste. A PUPA'S MOUTH IS ALWAYS CLEAN. Nest Mama was very vague. She was as vague as you could possibly be while still talking about something. Fard ãnd I figured out we were supposed to make shields of plaster of Paris, paint them Pupae blue, then stick letters from alphabet soup on them to spell the Creed of the Pupae. A PUPA IS NEW, A PUPA IS WHITE, A PUPA WRIGGLES BRAVELY. I hated this kind of gluey task, and so did Fard, the stuff would get out of control, you would get letters and junk all over your legs. But we poured our plaster and stood around. Jeans make you stand *around*. We were all wearing jeans except Nunzio in his regulation Pupae pants.

Look what she's doing, said Fard.

Nest Mama was *boiling* the alphabet macaroni of our Creed. She was a slave to package labels, she wasn't reading the instructions for our shields in the *Pupae Official Guide*. Look at the letters, what would the Phoenicians say. So all the Pupae were sad, there were no TREES to climb and our Creed was wet. But with the luck of all really mean redheads, Nunzio somehow got his letters stuck on his shield, probably the same favoritism that got him regulation Pupae pants. Now he paraded his shield around, holding his misspellings high above us, running at Julius with it, who started screaming.

I am finished, I am finished, chanted Nunzio, stomping in his regulation black Pupae shoes. He marched around Julius, whose shield was all messed up, it looked like a punched face. He'd started to paint it before it hardened so it was sea blue, not Pupae blue.

Nice blue, *Juliet*, said Nunzio.

It was the last straw, Julius raised his floppy shield and threw it but Nunzio ducked and the shield flew right for Nest Mama's REAR, she was bent over the table helping Gomez find a W.

Pow!

Nest Mama got a queer red face, she turned to her Sergeant-at-Arms. Who did that, Nunzio.

So now we all had to suffer Nunzio's great moment. His eyes got big like plates and all his molars shifted to the front of his grin.

Man, said Fard.

Julius! screamed Nunzio, come here!

But bravely Julius launched himself at Nunzio, who saw the Indian corn too late. The yellow and purple kernels intersected Nunzio's pale skin, brown freckles and red hair. Rejoice. Quickly though Nest Mama came over, what was left of the shield and letters dripping off her regulation Pupae skirt. She had to save her regulation blue plastic son from the corn wielder. She sent Julius back to his corner, where he was happy to be, though she didn't know that. He stuck his nose in the meet of the redwood boards and played weakly with the family's new-smelling garden hose. Nunzio had to be held back by Nest Mama and Gomez.

This could mean her epaulettes, said Fard.

We will now make Arts an Crafs boxes! yelled Nest Mama, ta keep ya Arts an Crafs materials in!

But she forgot that we had no materials, just wet lumps, and some of the third and fourth graders screamed out again. She held her hands to her ears, backed away from us, no, no, no more, this isn't, she'd spent so long painting her face and tailoring her skirt. She groped for the doorknob and escaped into her tiled kitchen. Nunzio stared at the door and turned to us, his legs apart and his hands behind his back, rocking on his heels like mean guys in the movies. *You made my mother cry!* he shouted, and ran inside.

We stood around for a while.

From the corner Julius said, Do you think we're going to get credit for this badge, even though we didn't finish our shield.

Don't bet on it, shet-ass, said Gomez, now suddenly blocking my way. Gomez was filled with martial spirit, you could tell he liked the way Nunzio had been acting. It made his eyes and mouth wet. You, he said. Don't forget the evil spirits are sending someone to pound your ugly mom.

*   *   *

141

I guess that meeting's over, said Fard.

That's what I liked about our street, my house was on the way to Fard's and his house was on the way to mine, depending.

I wish I hadn't paid my dues, said Fard.

I'm going to get out of this, I said, I don't care if I have to lie down in front of the street sweeper.

Or an ice-cream truck, said Fard.

Ha ha ha ha ha ha ha!

Once I decided something was not for me . . . You just KNOW. Dancing, German lessons, hot cereal—they all fell.

Gomez is stupid, said Fard.

Do you think he's really going to send someone to my house?

He does have his dad.

Well, I thought, maybe this is goodbye. I tried to take in everything about Fard, his house, for the last time. The washer and dryer seemed to get louder every week. Fard's brother lay alongside his scooter in the driveway like it was a girl. The garage was open and you could see boxes of stuff for the Broaster, not chickens, bags of BAGS.

It was great when you got hit by the ball today and Favorite Teacher laughed at you, said Fard, embarrassed. I don't mean it's funny you got hit, it was great to see Her laugh.

Yes, She is pretty when She laughs, I said. I guess this *was* a good day. Goodbye, Fard.

* * *

142

Our neighborhood seemed dry and stupid enough to snap off and blow away in the wind that was against me and my bike. At least I had a bike, even though it was small-wheeled. Another thing that made me weird, thanks a lot, was that tractor, my *spazz tractor* as Nunzio called it, which was all I had to ride until I got my small-wheeled bike. I liked the tractor for a while, you could put on Dad's gardening gloves and pretend it was the street sweeper, the guy drove with big gloves and a cigar. He looked like he had one eye, the way he watched the curb. He LEANED OUT.

All my stuff leads a stupid life, I thought, nothing's what it's *supposed* to be. My tractor was the street sweeper, or the *Disneyland—Alweg Monorail System*, I never pretended to FARM. You would have to drive the tractor back and forth across the backyard, Dad wouldn't like that. What they raised on farms were colors, rising and falling greens, black and white, red, east coast things. Julie and I had a plastic farm set but we hated it because the chickens had stupid plastic BASES which were even more stupidly supposed to look like the ground. Julius had a tractor, which was another reason I began to hate mine, and he had a disk plow you could attach. All he turned up with it were cat turds, they had a depressing back yard.

In kindergarten they took us to a farm. We walked to it from school, we were surprised to see it was right there. It's hard to see things that don't have neon signs. The animals, if there were any, mostly didn't come out of their barn and the farmer, if he really was one, didn't want to go get them. He seemed puzzled we were there. One pig came out then and rushed for Julius's shoe and sniffed it. All the way back to school we kept looking at Julius's shoe, like SNIFFING did something to it, it's ridiculous what you think when you're in kindergarten.

Things got tougher with the tractor after I decorated it with my NIXON LODGE sticker. That's when Nunzio started calling it a *spazz tractor*. Dad took me to see his Republican friend, this nervous guy who lived in an apartment. He had a white shirt so I assumed he was a Scientist and started talking to him about minerals, but then I saw he didn't know anything about it, he couldn't follow me. It was a short-sleeved white shirt. He gave me a big blue and orange sticker. *Those cost a lot of money*, Dad said in the car. He meant it was a nice gift but he didn't want me to use it, he didn't like our possessions to SAY anything. He yelled at a pimply kid who tried to wire a sign onto the bumper of the green car at Sea Lion Caves. *If you come one step closer with that thing, sonny—* I thought the colors of the sticker looked good on the tractor and I put NIXON LODGE on the front. Also I was the only guy on the block with a sticker.

That day was Army with Eric and Larry, who pushed me off the tractor and I watched NIXON LODGE on the nose of my TANK hurtle this way and that, attacking German positions. Ba-dow, ba-dow. The spazz tractor was indestructible so everyone wanted to get blown up on it. When the dads started to come home Larry gave it back. *Nixon Lodge is a stinker*, he said. My mom's for Kennedy anyway, I said. *Kennedy! God*, kid, *Kennedy* is a *shet*-ass.

Fard and I always went to buy *Mad* at the end of the month. We looked at the magazine rack at the variety store and our eyes went all over the place, from *Argosy* (a girl lying naked on a bed in a resinous shack except for a lot of twenty-dollar bills, looking at a guy in an ammo belt who's bursting in the door—Fard's dad left *Argosy* around the trailer but it was all words) to horoscope and puzzle magazines with pictures of ladies on them who look like the ladies who work for the Dental Borer. *The Old Farmer's Almanac*, I bought that sometimes, I kept it with my Herbert S. Zim books so I could look up tides on the east coast and think about them. With lighthouses and stuff. It always had this wisdom, these sayings you didn't get, *Potatoes meet potatoes when potatoes is done*, what is that, and then came a page of Pennsylvania Dutch jokes, like those people were stupider than the people who read *The Old Farmer's Almanac*. Fard usually rushed ahead of me shouting *The new Mad!*, we grabbed it, but I liked to save mine, I didn't like the way Fard tried to read the whole thing standing right there, showing me everything, his laugh of pain. Our favorite thing in *Mad* was Fidel Castro, he said *herk* instead of *jerk* and I liked his hat and beard. I liked them in the newspaper which I spread out on the floor to read the comics. The newspapers had thick headlines, I thought, that scare you from the driveway, we might be in a war. Dad came in while I was reading *Mad* in the tub. What kind of hat do you call this, I said, that Castro wears. *Fatigue cap*, Dad said. WHERE DO WE GET ONE?

I don't think that's a very good idea, Mom said.

Why not, said Dad.

Why, with all these terrible things going on—

What are you talking about, said Dad, it's not a *Castro hat*, that's just what *Joe* calls 'em. God damn it, it's a United States Army kind of hat and I guess the kid can wear one around here if he wants to.

The international situation was getting to Dad. Down we went to the old part of town, the Santa Fe station, its wig-wag, a big old hotel, the surplus store in the Moose Lodge. What's that?

That's where they—they just get together, that's all, said Dad.

Who? In the surplus store everything really had been in the War and it had dust and sweat on it. Army-green tins of talcum powder caked by moisture of the South Seas. After all, this is where Dad got his gardening boots. I was beginning to have doubts, being there made me feel like when everyone played Army and I was defenseless. But THERE WERE THE HATS, one of them even fit me. I plunked the tight fabric of the top of the hat with my finger and could not believe my luck, CASTRO HAT.

I could wear it driving the NIXON LODGE tractor, stick a Pink Owl in my mouth and be the street sweeper guy. LEANING OUT. Most people were going to think I was starting to play Army, I thought, that it was an Army hat, but it was a Castro hat and alternatively a street sweeper guy hat. I would tell them.

Grandfather came one Saturday and watched me driving the tractor around and around in the driveway like a moron, with the hat and Pink Owl.

Are you planning to join the Army? he said.

It's not an *Army* hat, I said, mad, and saw Dad drop his tools and come running toward us really fast as he heard me, it's a *Fidel Castro* hat.

Grandfather was laughing the I-don't-think-I-heard-that laugh.

Dad butted in, He got the *idea* from Castro, he wants to look like Castro, I don't know, beats me (Dad started to look like he was feeling funny saying this out loud in the front yard), but it's a regulation hat, Pop, you know that.

Well I, said Grandfather, if that's the reason he's wearing it, he—that's the wrong reason.

The three generations of us argued about my hat there on the driveway for an hour. It was a US Army hat but I was wearing it in partial imitation of our ENEMY. But I had a Nixon tractor.

He's just playing, Pop. What are you playing?

I'm the street sweeper, here is my Pink Owl.

We couldn't figure it out.

The word *play* made Dad mad because there was

147

a lot of important stuff on tv about the REDS. I went in when he was watching tv about that time, Nixon Lodge time, and he gave me and my stick horse, which I just happened to be using, I hadn't played with it in a long time, a really mean look.

This affects you, he said.

You mean I should watch this?

It's an idea, buster.

But this had the OTHER TV look to it, the stuff you couldn't stand to watch, it was way beyond Bill Welch. So grey, those things didn't make sense, they sounded hollow and they didn't move as if they were alive. The suits they wore on that kind of tv looked just like their haircuts, like the screens on the microphones, like the textures of their voices, rough like the vinyl couch that ate into your legs while you were trying to watch.

Here was Mr Neighbor, raking, he raked all the time, even if you just saw him out of the corner of your eye he made movements like raking. On weekends you heard the engines of little planes, and small blimps sailed around the county. Some sounds came only on Saturday and some on Sunday. Mr Neighbor had these teenagers, Jim and Jimmy, with model planes hanging from the ceiling of their room. Up in the rafters of the Neighbors' garage was a big gasoline model sea plane, and a rich wood radio-controlled yacht. Mr Neighbor had a black and white crew cut. He always wore blue canvas slip-on shoes

148

with gum rubber soles, pants from somewhere between suit and boating, a shirt of plaid flannel. He smoked a drugstore pipe full of bright and burley, when a feller needs a friend. Mr Neighbor always looked up and said hello or waved if you did anything in the driveway. He never stopped raking to talk and he only put down his rake to go inside and have a Monterey Jack cheese sandwich. When he said *hello* or *hello boys* it was in a voice made of his smoke, the light, the leaves, his shirt, it was a plaid voice. If you looked down his throat with a flashlight you'd see a vibrating three-dimensional plaid of these things, it was on my MIND DIAL.

On the weekends, the small engines, Mr Neighbor raking, the lawnmowers, blimps, the Thimbledrome models at the schoolyard, there was something in all the leady gasoline smell of the past, of the WAR, little pools of the War drying up in our county. There were grey ships in San Diego. The dust that accumulated on 78s, the tiny smell of warmth the glowing radio tubes contributed to our homes of hooked rugs, frontier lamps, colonial sofas. DER BINGLE could have made himself mighty comfortable by that combination reading lamp and pipe stand.

Even though we didn't play Army much, Fard and I knew the War was real because it was strong, bright, and slow-moving on tv. Saturday afternoons I watched war movies with Fard and his dad. Fard liked burning airplanes but you could

149

tell he was drawn into the War by the look in his dad's eye, his practically folded-away eye. Fard's dad *liked* the talking, we sat there out of respect even though it drove you crazy. You sat there in the smell of laundry and wondered why the War seemed closer in Fard's house, was that where the chaos began. Who DIED.

Oh yeah? I'll tell you one thing, Sergeant Gunderson.

What's that, *Captain?*

The krauts have a hell of a lot of guns emplaced along that river. And you know what, Sergeant?

No, *what?*

Each one of those Natzy guns has a name. Not Fritz, or Adolph, or Eva, *Sergeant*, or Grösse Palooka, or anything like that. Those guns have names on 'em like Tommy, and Mary Lou, and Jimmy Brown the Newsboy. Or Old Pop, *Old Pop*, Sergeant, who runs the gas station and makes wreaths for the orphanage at Christmas. You know him.

Know him, why I—

That's enough, Mister! [Slaps him.] So don't go telling MY MEN it doesn't matter if we knock out those guns before we get to the target. We don't hit those guns tonight, no more village green, no more Chestnut Street, no more little Sally the—

Are you all right? [Slaps *him*.]

Just do your job, soldier, and I'll be waiting right here when you—

Look at that, Fard's dad said through his loamy

grin, pointing with his huge hand and its beer can, that's how to tell 'em.

The War was a lot more interesting to Fard's dad than the Broaster and driving their red and white truck.

Dad tried to ignore the War. To Dad the War in movies was devious ensigns fattening chickens below decks, visiting hula girls until very late at night, that's not how it was. *Is that beer I smell on your breath, Mister*, bah.

They never talked about the War at school, our school wasn't even there during the War. Favorite Teacher looked like She might have spent the War waiting for something, Her handsome father to come home to Her beautiful mother, Fard said. If you wanted war you played Army, watched movies with Fard, or made models. The models Jim and Jimmy hung from their ceiling seemed grim, like the War, not like the models in the variety store now.

I found out Jim and Jimmy watched MY MOVIE, *Run Silent, Run Deep*, that Dad showed on the dining room window shade for my birthday party. He rented a projector from the library and let me draw the curtain back slowly when he showed the cartoon, just like the THEATER. Mickey Mouse wasn't funny, Fard didn't think so either, nobody thought so, but the TITLE of the cartoon was red and soft, like pillows in the dark. *Run Silent, Run Deep* was dark blue and you only got to see the submarine between the TALKING. I asked Dad if

151

there were any squids in that movie, not giant ones but just any. *Don't think so*, he whispered. All the guys on the submarine talked louder and louder, they hated each other more than the JAPS. *Let me tell you something, Mister, I'm the skipper of this coffin and—*

Whew, said Fard, how come these guys are on a REAL SUBMARINE and they just talk?

Jim and Jimmy tried to get tough with me about watching the movie from their bedroom window, like they'd gotten away with something for FREE.

Hey, we saw your movie last night.

So?

It was really stupid, *kid*.

That's cause you couldn't hear the talking, I said, you guys are so dumb. I couldn't really say that to them, *you are dumb*, but I wanted to. Next time, I thought, you'll have to strain your eyes to see *Run Silent, Run Deep* because our honeysuckle grows in front of the window and you can't even prune it because it's full of BEES.

Our driveway, the pieces of shells in it. I could get all the way out here with my Bill Welch microphone. When it was hot you could run through the sprinkler and lie down on the driveway and leave a wet skeleton. One really hot day I did another thing on the driveway which made me seem weird to everyone in the neighborhood who hadn't bothered about it up till that point. Thanks a lot. It was hot, clear and still and the whole block was quiet,

152

everyone was at swimming lessons or on vacation or hiding inside with their air conditioners. The perfect day to sell lemonade at the end of the driveway. I pictured everything, the folding metal tray, my SIGN, the waxed paper cups, except there were no waxed paper cups and no drinks in the refrigerator to sell. Only orange juice and nobody would pay money for that. They hate it. Up on a high shelf I found tiny glasses with stems, and in the cupboard I saw a can of evaporated milk, which I'd never tasted, but milk, I liked cold milk on a hot day just as much as anything Dad kept yelling was poison. I made my sign and got out there, right in the middle of the driveway, MILK 5¢, with my four fancy glasses and my can of evaporated milk. The neighborhood was an even greater blank than usual. The few people who came by looked at the sign and gave me the orange juice look. Only worse. A guy in a hat walked along and stood looking at me, I think he was a salesman. Milk, I said, five cents. *No thanks*, he said. After I sat there for an hour, during which of course no one told me that what I was doing was *DUMB*, I started to get really hot myself, I needed a drink of refreshing evaporated milk, a little warm now in its can but looking delicious and—yellow—in—

Ran churning into the house and everyone had been watching me from behind their window shades to see me finally taste my wares, as it turned out.

I put my bike in the garage. I couldn't remember

which side I hurt when I Fell Off My Bike, so I could tell Dad, wait until he heard about Pupae, I thought, remembering Dad with a shock in the dusty, metallic, lumbry garage that came alive with him on Saturdays. Saturday was a day of torture because of our heritage of thousands of second-hand screws in baby food jars. And Grandfather's tools, worn but strong with black or green or maroon wooden handles. There is always something to do at the work bench on Saturday, a plug to be replaced, something small to be made out of wood, our heritage included the making of shims. THINGS NEED TO BE BRACED. Grandfather continually made very ugly things, step stools, unexplainable racks, boxes, carry-alls. But BOY were they sturdy. He kept bringing them over.

Dad didn't like the chores of carpentry. Or gardening. Working for hours with his hands bored him and made him cranky, even though he inherited Grandfather's talent for making triangular corner braces. When it was time to brace or fix something you had to be doing it NOW, there was no escape, you were a PRISONER IN THE GARAGE until the thing was marked, sawn, sanded, glued, nailed. Varnished. You couldn't GO anywhere else. Dad liked to THINK, he likes freedom, like I do, I thought.

When he had his work cap on his head, his safety glasses on, his work gloves ready, under the work light at the work bench, in all that nail smell, he told me about the Big Companies. He had a lot of worn cardboard tubes, DuPont, Dumont, Dunlap, Dunlop, Dulux, Delux, Duco. Companies that liked to put their names in ovals. While he used the stuff in the tubes he told me about the companies, like a history lesson, as if they were NATIONS to him. *The full name of this company is E. I. du Pont de Nemours. It's in Delaware. They began by making gunpowder . . . RCA stands for the Radio Corporation of America, do you see the lightning bolt under the A . . . Dunlop makes a wide variety of glues and polymers . . .* The smell of the giant tool chest, rusty dust, solder. When he soldered Dad sometimes got emotional about the Big Companies. *They brought us through the War . . .* He looked at the names in the ovals on the tubes and thought about things, then he'd tell me about the *Mighty Boy Builder*. When he was growing up in the north woods there was a shelf, carpentered by his Pop, next to Dad's bed. He had nine blankets and a thick olive-green comforter, Julie and I knew all about his bedding because he always started *describing* it when Julie begged him to turn up the air conditioner, she got hot. He *often awoke in youth to find his bed covered with drifting snow*, which made her cry with frustration but also pity. But, he always said, he was *perfectly warm down in the heart of the bed*. On his shelf was the *Mighty Boy Builder*, a book he got for Christmas. Every night Dad told himself he was going to look through the whole *Mighty Boy Builder* before he fell asleep, but woke every morning with his nose pinched in page 35, BUILD YOUR OWN CANOE. Half the projects in the *Mighty Boy Builder* were to be made from soldiers' helmets left over from World War I, not the canoe, but. There weren't a lot of helmets in the north woods, though Dad and his friend once found ten of them in his friend's uncle's woodshed. That Christmas everyone received helmet reading lamps, ashtrays, floral vases. AN INEXPENSIVE AND FUNCTIONAL CHICK INCUBATOR FROM STANDARD SOLDIERS' HELMETS. You could make almost anything in the *Mighty Boy Builder* if you had a helmet, old wheels, and a crate. Sometimes I dreamt of all the stuff lying around in the north woods for Dad to use, half-done spools of wire, slats, black enamel doorknobs, hinges, sawhorses, paraffin, tin, blotting paper, kerosene, railroad spikes, KEGS. CHEESE-GRATE AND ASHTRAY MADE FROM A TIN CAN. A SMALL WORKING PILE DRIVER. BLOTTER ATTACHED TO WRIST SAVES TIME. SQUIRREL-SKIN BILLFOLD. Looking around the garage I thought I liked to make things like that, too, not just my MACHINES that drove Dad crazy, my *unconscionable waste of nails*.

I'm never going THERE again, I'll tell you that, I practiced saying to Mom, while I was putting my bike away in the garage. I wanted it to sound like Dad talking. I made my voice echo off the cement floor, up into the rafters, where I always wanted to live, where you could smell wood and hear the 'rain'. Of course she was going to ask me about the Nest. When I got in the house though she had a kind of funny look on her face and instead of telling me all the reasonable reasons why *that is not the case*, she asked me if I wanted a honey sandwich. Okay I said. Hmm I thought. *Julie's at Kelly's*, she said. I went to wait for Julie in her room, to get ready for the animal tea party and ROCKET TRIP.

Here's Julie. Seven, tan, we both went around without shirts or shoes most of the year. She had a strange hairdo with flips, this way and that, it kind of made her look like a dog, I mean it gave a lot of lift to her smile and optimism to her face. Everything was parted, moustache-shaped, with her. Big grey eyes with which she shamed adults, she *accused* them in just looking at their flat boring faces. She stumped around on short legs. Weird Shorts for Girls. Red socks in white sandals. She had a weird way of walking, she appeared always to be following but was leading. She hung around only with Kelly, and with her own cat, Greyie.

I never knew what the right thing was, I mean I kept losing track of it, but Julie always knew and she eliminated our neighbors regularly with her burning glances. Dad was afraid of half of our neighbors because they had got the treatment from Julie, for yelling at their pets or having a messy front yard.

Once a month I dreamt we ate dinner in the dining room, at our usual places, four big red

157

candles in the middle of the table, our eyes glinted. Julie reached out for a glass of milk and knocked over one of the candles, then knocked the milk over too. The milk snuffed the candle and mingled with the red wax, which sent a SIGNAL up to the ceiling for thousands and thousands of BUGS to descend on us, like fine rain, large and green and brown, they kicked and tickled, but not as though they were completely there—you felt and did not feel them. The house filled up to a depth of several feet with bugs, falling down into them seemed to be the thing to avoid. I lost sight of Dad and Julie and Mom was off-balance . . .

Julie and Kelly ran up and down the sidewalk with big scarves tied around their necks, playing Mighty Mouse or Mighty Mouse With Kelly. Sometimes they were Mighty Mouse simultaneously and sometimes Julie was Mighty Mouse and Kelly was Kelly. Kelly's brother never stopped moving, he was always bugging you and trying to hit you or push you down. I had to keep protecting Julie from him because she might skin her knee. Julie and Kelly's sister didn't get along either, Julie BUGGED everyone at the Kellys', she was too smart for them, and too scared at the same time. The Kellys had lots of scary stuff that was broken and weird, but you ended up having to use it anyway. They had a *plastic slide* in the back yard that was cracked, when you slid down it opened up in the middle and you either fell out or it closed up and pinched your bottom really hard. They also had a thing you held

on to and slid along a cable, which ought to have
been fun because it was like a RIDE, we wanted
everything to be like a RIDE, but the Kellys' fat dad
could only stick the cable part way up the pole (he
was using the telephone pole on the other side of
the fence which could get you in TROUBLE) and he
couldn't tighten it with his big stomach straining
so the cable boinged way low when you slid along it
and then you caught your leg on Fort Dogpooh and
it dumped you right in there.

Fort Dogpooh was a huge thing the Kellys were
a-building in their back yard, year by year, Fort
Dogpooh was my name for it, not theirs. It was
made of old blankets, card tables, tarpaulins, parts
of fences, wavy fiberglass, refrigerator boxes. It was
like the Casbah, I bet it had fifty rooms to it, every
one filled with dog pooh. As soon as their stinking
dog, *Brad Kelly*, poohed in one room of the fort,
Kelly's brother got some junk and added a room.
Some of the rooms with colors were fun to be in
on a hot day, except for the smell, or were cozy
in a 'rain storm', but it always smelled like pooh.
But the Kellys weren't frightened of dog pooh,
they just walked on it like it was regular dirt. No
wonder they had no carpets in their house. But it
was outrageous, the fort was really neat and they
were building it for *Brad*, for the DOG. I had dreams
where I roamed Fort Dogpooh, wonderful things
went on in there, there were rooms with waterfalls,
rooms with RIDES.

This is how scared Julie could be: at Disneyland

we were on the pirate ship, there was a guy with *really no leg* whacking around, brandishing a cutlass and showing off his ornate wooden leg from a piano, talking like they do, *Avast, blah blah blah*, Julie got a load of this guy and grabbed me. *That's not a real pirate is it*, she asked me. What are you talking about, I said, there is no such thing as a real pirate. The guy whirled around on the Steinway or Stuyvesant leg, *WHO SAYS I'M NOT A REAL PIRATE!*, glowering at Julie who looked like she was going to go off the edge of the boat. Dad pulled us down the gangplank and started lecturing me, I was surprised, *Listen, pirates were terribly real until very recently*. I thought they were like cowboys. *They are!*

But this is how brave Julie was: one day she got mad at Kelly's brother and sister, they'd been bugging her and Kelly while they were trying to RESCUE someone, so while the Kellys were in their house eating the horrible lunch they always ate out of cans, Julie went into their garage to look for something to GET them with. She was small but she picked up a whole box of bottles, bottles of root beer, and took it into Fort Dogpooh. She crawled all around in there, dragging the box of bottles behind her, through Brad's poohs. She left it in a room that looked like no one had been in it for a while, where the pooh was white. The Kellys were madly cheap Baptists, they made their own root beer from little bottles of extract, *Yeow! What's the point!* Fard yelled when Mrs Kelly 'let' him taste some of it on Hallowe'en. Kelly's brother got a real beating from their fat dad when he couldn't find the bottles. Good. Punishment was a big deal for the Kellys, they had wooden paddles up on the wall like decorations, paddles that said MAMA'S LITTLE HELPER and FATHER KNOWS BEST and PADDLIN' MADELEINE HOME. They also had a GIRLS bathroom and a BOYS bathroom, like at school, which I thought was kind of neat, in a smelly Kelly way. *That's the stupidest thing I've ever heard of*, said Dad, *why in THIS family we have a GROWNUPS bathroom and you have YOUR bathroom. The PINK bathroom and the GREEN bathroom.*

Months later Kelly's brother was crawling around in the fort and he found the box of bottles. He dragged it out in triumph, covering it with dog pooh, but the Kellys didn't care about that, they just wanted to get their mitts on all this glorious root beer they never thought they'd see again. They all started drinking it and it had fermented, Dad said, and then it was late at night and the Kellys' lights streamed out all over the block and scary noises came from their house, singing and crying and yelling and everybody getting beat up. Dad and Mr Postum went over to investigate. In the dark they saw Gomez's mom in a woo hoo nightie, peering over at the Kellys'. Their fat dad stumbled out of the house into the driveway just as Dad and Mr Postum were walking up. He stood there swinging their cat, Ludwig, around his head by the tail, saying, *Here kitty kitty, here kitty kitty.*

* * *

Julie wouldn't mind if I went in her room, not automatically. She wasn't protective of her stuff. Besides, the dolls and pink things had nothing to fear from me, I hated their guts. There were a lot of smells in Julie's room. She had play perfume you bought at the supermarket (I had a shaving set with cardboard blades, Mom had to get us the same kind of stuff or it was WAR) and Julie put that perfume on her toys. It was watery and smelled like bubble bath. She treated her toys so well. Mine I threw onto the roof, set on fire, or put helpless in front of the street sweeper. Some of the dolls were perfumed, they came that way, I bet there is perfume you can buy for dolls too, they DEMAND EVERYTHING. There was modeling clay in the smell, greasy, in swirled colors. Fuzzy lumps of clay were all over Julie's room, little balls and cubes of it in most of her drawers, with the play food, carrots and corn on the cob and fried eggs in tiny metal pans. There was also the smell of raw erasers, rubber she chewed and slobbered on, strong-smelling plastic Easter eggs, the wet lawn outside, the little bowls of cereal and milk Julie always ate, and soap.

Her hat and her gun wuz a-hangin' up on the coat rack, she was the only cowboy on the block. We were both given cowboy outfits, at the same time, naturally, to avoid murder, but my mineral hammer gave me a much mightier feeling than any gun. Than any tin star, stranger. On a hot day Julie went around the neighborhood on her tricycle, in

162

her red cowboy hat, boots, holster, with only her weird worn-out pajama bottoms on. She watched cowboy shows on tv but I didn't, I left the room, they're so unScientific. They look hot and dusty and uncomfortable, not heroic. I hate that *muffled noise* of the horses' hooves when they gallop on soft earth, which is right around here anyway, it's not the *Old West*, how dumb. Cowboys seem stupid, they always have to catch someone. They're always *bursting in* when it's too late. Their gun belts look stupid. There's never anything exciting at the end, it's always just *Say, where's that roan got to, Pablo?* And *Well, okay, thanks a heap, ma'am*, they don't even say *pardner* like you thought I was a-goin' to there. The way they talk is stupid, it's like waiting for Grandfather to finish a sentence. So SLOW. The bad guys have to share the same moustache, the furniture is ugly, you can see airplanes and telephone wires all over the place. They treat Mexicans like crap, they treat Indians like crap, they treat farmers like crap, they treat horses like crap, cattle like crap, girls crap.

I yelled about this to Fard and he said, *the only thing cowboys like is their own pants*. The music is really dumb, *buh! budda buh!* When the camera's alongside a stagecoach the wheels do that thing where they look like they're running backwards and it makes you SICK. There's really stupid kissing, the cowboy grabs the girl by the upper arms, to show he's worthy even though he just licked her nose and mouth without warning, and then she has to look up at him and say *Oh Jimmy, do you think you CAN save Dad's cattle?* He only gets another kiss if he does a stupid cowboy *chore*.

They sing, suddenly in much nicer clothes than they just had on, stupid songs like *I'm Thinkin' I'm Anglin' After Hankerin' for a Filly Like Yew*, why don't they just go GET them? Haven't they learned anything from Popeye? There's a BAD GUY at the Bank who's going to get the girl, in exchange for shares in the Railroad. So what. They wear bandannas. They only wear them in these crap cowboy towns, you don't see people in New York in 1890 wearing bandannas. It doesn't matter where you put a thing like that, it makes you look *STUPID*.

Nobody played Cowboys on our block, it was all Army or Wizard of Oz. Cowboys weren't real. At a motel in the desert Julie shot me over and over again and I fell backwards into the pool. Every time she shot me she yelled out *Take that, you dirty son of a bitch*, until we suddenly had to go upstairs. I'd rather watch a month of war movies *with talking* than the stupid cowboys ever again.

Maybe Julie liked cowboys because she got to watch a lot of animals on a cowboy show. That's what she was after, ANIMALS. Why do girls go for these animals, I asked Fard, do they know something? Anita and her horses, Roxie and her rat, Julie and her toy box, her *pencil* box was full of animals, she had big collections of tiny rubber and plastic animals besides big STUFFED animals. But take her to see the seals and she's scared of the waxed paper bags of fish they sell you. Take her to the aquarium and she's too frightened to walk around the ocean tank. Take her to see the baby elephants and the next thing she's screaming because the elephant *blew its trunk* on her hand in which she was holding out a peanut. *Everyone blows their nose, sweetie, a lot, don't you blow your nose a lot?* Dad said while taking her to the wash-hands place with her big handful of *trunk snot*. Dad is not comforting, he only offers you reality. One dark day we went to the kiddie zoo, so right there Julie was all emotional, and a mule lives there, not in a barn, just a kind of car port with a manger in it. 'Clouds' were gathering and the zoo was closing as we stopped to look at the mule in his little booth. He had a sad ironical mouth, they all have it, mules, burros. Julie looked at the 'rain cloud' and asked Dad *Where does he sleep when it rains?* And the mule broke out in loud gulping SOBS, not hee-haw, a devastating cry from the whole history of donkeys. Julie didn't stop crying until after dinner.

Her toy box was a riot of anatomy, flesh-colored plastic, adult and child limbs, big fat cloth legs of bright colored happy animals. The box smelled of the worn cloth of the limbs and the strong plastic of the dolls. I looked down into the box and it looked like a FIGHT. The dirty yellow rabbit Julie loves more than anything was wearing my glow-in-the-dark tiki around its neck. The pathetic rabbit comforting himself in the dark.

165

TIKIS are so important we don't really know what they're for. *They have something to do with teenagers*, Fard said. He had a black one with white stripes and eyes. In the light, my glow-in-the-dark one was pale green with rhinestone eyes. A couple of years ago, Julie and I decided we would only get glow-in-the-dark stuff. That didn't last long. At a carnival in front of the supermarket she got a big plaster cat that glowed purple but only for five minutes. I got a glow-in-the-dark rubber skeleton, and we had some glow-in-the-dark counters from a game, all in all it wasn't much of a collection. For a while she kept all our glow-in-the-dark stuff on a shelf in her glass cupboard, the skyscraper of the rubber animals, then she didn't. Here she was.

Time for the party, she said.

And rocket trip through inky blackness. Julie, what animals are you going to bring, and are the dolls coming, because—

Well, I'm going to invite them, said Julie, but who knows if they can come?

Where's it going to be?

Let's say it's in a big garden. They live in the skyscraper.

She started pulling stuffed animals out of the toy box, rubber animals out of drawers, animals animals animals, they joined us at the tea table, tiny animals in saucers, around the tea cups, big animals got chairs or climbed on one another's shoulders. How could the plastic-faced ponies and Lambert the enormous stuffed sheep have been

166

interested in such small cups of balloon tea? Julie made tea by putting used balloons into the tea pot, red, blue and yellow, or around Hallowe'en, orange and black.

Let's say it's a rocket, not the skyscraper, and we're in it right now. Already.

Okay, she said, they're having balloon tea right before they blast off.

Or they're having balloon tea with special medicine that makes it so they can breathe without helmets.

Yeah!

I did the blasting-off noises, to show what I'd learned from Nunzio, he never did just *T minus two minutes and counting*, he knew all this other stuff, what happened at thirty seconds before lift-off, eighteen seconds, that the engines lit five seconds before zero, Julie was impressed. But she wanted to say it, too, which made me a little mad because *I* was saying it, she didn't have a sprayer. We pulled down the window shade, now the tea party was ANTI-GRAVITY. The animals floated and drank balloon tea out of floating cups and saucers. Lambert started bossing the scraggly yellow rabbit around.

What's the idea, said Julie.

I told her tempers were fraying. It was like on a submarine, it was war.

Let me tell you something, *Mister*, said Lambert. I'm the skipper of this tub, and this is a stuffed animal rocket trip—

—and tea party, said Julie—

167

—correct, said Lambert, and *no dolls* are coming on this mission. Dolls use up WAY too much air.

Julie would go anywhere, I could have said it was the inky blackness of Neptune's realm and it would have been fine with her, as long as she could look after the animals and give them balloon tea. She was bold, Julie, she didn't need her cape on to be bold and face danger.

More balloon tea, Sir? Looks like it might be a little rough up ahead. How about some more metal ears of corn for the men, Sir?

Get a grip on yourself, lieutenant, said Capt. Lambert.

I heard all that, said Dad suddenly from down the hall. The Navy was not like that. It wasn't like that at all. I'm home.

Cripe, I thought. I had to get my story straight about the Nest, I had to sort out which side of me had the heroic injury which must keep me out of the Nest for weeks, *barely able to attend school*.

The harsh shadows of plates being passed over the bumpy plastic tablecloth like rhino skin, we had to have dinner under the glaring frontier light at the kitchen table. This meant something was up. There were going to be DISCUSSIONS. Julie looked back and forth between us.

Dad cut his chop Scientifically. He talked to me but looked at Mom, who was dainty and ate her chop with a secret smile.

Did you see Gomez at school today? And did

168

you tell him *very specifically* what I said, about the language you committed against your sister and me?

I realized from now on I had to say GUY like the little kids instead of GOD, which leaves you, as Fard has pointed out, with the ridiculous GUY YOU GUYS.

Guy you guys, Gomez and Larry made us go to the Big House after school and they had evil spirits in the ground, Gomez said they were mad at you. They're going to send someone to POUND you and Mom.

Larry who? This is ridiculous. Did he really say that? Someone is coming?

Yeah.

Gee, that constitutes a threat, in law. I could call the police.

Now now, said Mom, can't we just—

Dad got up and started taking our plates to the counter, even though we weren't finished. He turned on lights we never used, over the sink, in the hood of the stove, symbols of torment, the HOME is under ATTACK. He washed the dishes, banging them against the sides of the sink. Julie began to cry. Dad thought about our neighborhood, what Mom told him about it, the other moms wore zoris, they smoked, they drank coffee *all day long*. And Walt Disney had sent us there. Dad didn't have to put up with any of it, except maybe Jack Hass, his beercan-shaped head, the loudness of guys with those-shaped heads. But Dad had seen Gomez

Grande crouching on his poster. *Smog, illiteracy, biceps*. In the soapy water Dad absently felt the long blade of a knife and realized he'd cut himself. God damn it anyway. He went out, to the bathroom, turning on every light along the way.

After dinner Dad put our whole family on the sofa. He turned it so that it faced the front door. All the lights were on. *No* to tv, popcorn, toys on the sofa, fidgeting, a story.

I waited a long time and then, like usual, I forgot there was a lot of tension, forgot Dad was expecting the evil spirits at any moment. Sing 'Sweet Betsy from Pike', I said.

No!

He and Mom talked a little. He looked hard at the paper, as if looking for news of himself and the evil spirits. The REDS were doing something. Maybe they were the evil spirits, I thought. Mom twisted her hankie and sat like she did in church.

Hey, he said suddenly, how was it, how was *your first day at the Nest?*

He hunched forward, trying to get away from what might be going to happen to us. I knew what I had to tell him, to RESCUE myself. All the forests he was from, the miles and miles of pines, having to snowshoe to his miserable one-room school which wasn't even red. I knew he saw my life as strangled, *empty of woodcraft*. So I told him how we had been DENIED TREE CLIMBING by the Captains of Pupae, who live in *Bellflower*. In all our cement

170

neighborhood there was not a TREE for a BOY to CLIMB, under our shellacked Nest Mama I was not going to learn the THINGS A BOY MUST LEARN TO BE A BOY. I really laid it on. I wish Fard had been there. Dad looked grief-stricken and terrified, he kept looking at the front door, through it, out into the past of America, as if you could see that from there, the north woods. Waiting for the evil spirits. He turned to Mom.

This boy, he said, this boy is no longer a Pupa. Do you hear.

It's a ruse, Mom said, you should make him go, he Fell Off His Bike.

No, Dad said, he must not go.

He got up and went to look out the window in the front door. *No trees*, he said. I will take him in hand.

Pretty ominous, I thought, what does he—

Nosebleed. Mom brought me cocoa in a glass mug, white like Tutankhamun's alabaster head with the red nostrils. The cocoa went down sweet and slimy like blood, but I was replenishing my fluids, Dad said from where he was looking out the front door, that's what the Egyptians always did, Joe boy, replenished their fluids. And manfully and mercifully now he let us escape the sofa, let Julie and me take ourselves into the cornucopia of the dark.

171

Fard called it the bullseye but it never reminded me of that. From the first time, I saw the flying WB, the concentric rings, clearly as the mouth of a cornucopia. In you go. From behind a hill beyond a barn, the classical farm windmill in deep purple—not black—silhouette, and the rangy low fence: dawn, and the dawn music from *William Tell*. Here were hilly fields of warm green, a collection of greens from story books, school books, posters, brochures. The loam was rich, billowingly plowed. The rooster crowed at a precise angle to the sun, as we knew he would. The rising sun picked out the clean red of the tractor, immaculate milk cans, the red of the barn wall. The sun rose faster now as the narrator had almost finished his text. Things were never like these colors, but our nation wanted them to be.

The farmer's hat was gold straw. His overalls moved not like denim but liquid indigo, his red and white checked shirt against the sandy farmyard floor, the early morning sky, TREES and fields in the distance. Melody of the sleepily awakening oboe. In the henhouse cloud-like hens mimicked

172

the factories over the hill, eggs rolling from their fannies every few seconds, rolled elliptic and funny down chutes like drain pipes. Production would increase with the introduction of Raymond Scott's 'Powerhouse'. Were there Marching Eggs in cave paintings? Mosaic ova with yellow-orange legs on Roman walls? Eggs marched along, one got lost, one was comically drunk, one belligerent, sometimes they hatched. In the black and white days they never hatched, when it was Fletcher Henderson instead of Rossini.

In the field the farmer plowed, distant on his tractor, perhaps *Turkey in the Straw* as he softly undulated. Pesky gophers pulled carrots out of the foreground into their tunnels, flooded later by the brutal farmer, Julie crying out, though this water was like blue paint. The carrots disappeared rapidly in rows with the sound of a pop gun.

The barnyard, Grand Ole Opry *buffa*, rivalry of rooster and dog, the plaintive love of the little old hen with spectacles for the big manly fellow. Once on a stage here the Sinatra rooster and the Bing Crosby rooster competed for the affections of swooning bobby soxer hens. How else would Julie and I have known what a silo is, what a pump is for, a churn? This is where our food came from. Whenever I saw the egg lady on our block I hummed the dawn music from *William Tell*. Even though they existed in our semidesert, no one could tell us what the fan-bladed windmills were for. We found out in the Technicolor farmyard.

At the barnyard gate there was a mailbox, little Quonset hut with its red metal flag. Beyond the gate there was the road, which called, as though its paintbrush had burned with love through this landscape. In 1935 you could love the open road, drawing it in your hot little office in Hollywood, because the road *was* these colors. In the dark Julie and I learned them, we looked for them when we traveled in the car. The road had changed but we looked out the windows like they were tvs.

On the road the Rabbit and the tortoise raced, sometimes classically, or with help, sometimes dressed as each other. They raced by fences and green hills, raced the scenery itself which came again and again. A million tortoises behind the scenes. The Rabbit's friends beat to a pulp the fake shell-wearer. I wonder if there coulda been more than one of them little guys? *Mmm—could be!*

On the road the city wolf traded places with the country wolf. The country wolf couldn't behave himself in nightclubs (this was during the War). The farm jalopy sped toward the city, the nightclub, the girl, VA-VOOM. The city wolf's low speedster stopped and gleamed in the dust just long enough for him to go ape over the belle of the barnyard, his eyeballs telescoping out into red and white casaba melons with rhythmic klaxoning, AOOGA!, his tongue a red carpet unrolled for several hundred yards and already walked on by the many dusty feet of the barnyard, her beauty ungathered by the hick wolf for her milkmaid's bonnet.

Most of all on the road it was the mailmen. On scooters and in trucks they raced over the very humpy hills, the hills of BUELLTON I thought, that was a green place we were once, though they were meant for giant, everywhere-Ohio. The mailmen appeared, disappeared and appeared again at top speed, hats whipped backwards, delivering telegrams, instantly bringing stinging replies (*P.U.!!*), cakes, notices from the draft board, put-putting along importantly under the clouds of war. The War, when the Duck tested pregnant shells with a ball peen hammer. It was something funny happening in your house at night and suddenly from the street the cry PUT OUT THAT LIGHT! or midway through your pursuit of the Rabbit he turned around, suddenly in a warden's helmet and roared you down, his mouth a gaping lion's, *QUI-ET!* Or suddenly demure, *Is this trip really necessary?*, then smashed you on the head with an outsize mallet. A fairground strong-guy bell-thing mallet, as Julie would say. We drove Dad crazy by calling the human uvula the *punching bag*, since the only use we knew for it was its punishment by miniature boxers in the throats of our heroes.

But the main use of the mailmen: you are chasing the Rabbit though he is far ahead of you. He has time to reach inside his fur and grab out pencil, paper, envelope and stamp. Hurriedly he writes—you're still coming—licks the stamp, the

175

flap, posts the letter and leans against the mailbox whistling, examining his nails, vaudeville, as you come closer to him a mail truck passes you, driver drops enormous package into the Rabbit's arms—*Package for Bugs Bunny!*—with one pull he undoes the string and brown paper and is delighted to find an enormous pie which he RAMS into your face and entire upper torso just as you are upon him, your hands clutching out to GET him but useless now, impossibly slow under great slabs of cherry filling. If it was a really bad pie it oozed down your head, a few cherries doing a quick impression of your own face.

If you were chasing him around a restaurant the Rabbit baked a cake himself, barely finishing the ornate decoration of it before at the very last second tipping it up to receive your anguished point-of-no-return face. Then Humphrey Bogart, who was in the dining room with Betty, laughed at you. Then he threatened you.

But all that is in the city where the road led. Slurpy string quartet arrangement of 'Oh, Susannah': the road led too to the Wild West, to the War, to America's playgrounds, in the peculiar watercolor box of postcards. Now it led over a hill: a pond, stretched large and small, plumbed ocean-like, curiously frozen or thawed at the wrong moments, year in, year out. Dawn and *William Tell* came to the pond again, cat tails like corn dogs. Where the blue of the night met the gold of Mr Schlesinger's day, ducks in a V. For Victory—we knew that. Real

176

ducks, not at all like the Duck about to march out of the rushes and slap down Fudd's gun, slap it down and berate him, insulting, *scathing*, mad. *Gee, I'm sowwy, Mr Duck, I didn't wealize*—Fudd in his silly tall hat, helpless, one blast from his idiotic trumpet-shaped gun sent his boat half way back across the pond or into the drink, the bottom of the pond connected by tunnels to all the underwater places. Sunken galleons, laughing octopi, giant clams yawning, the Cat bubbling in diving helmet and lead shoes in search of fish, the Rabbit dressed as innumerable mermaids out of the inky blackness of Dorothy Lamour's wildest closet. For Fudd the pond was a kind of glasshouse, a fun-house of threat, nose-honk and hat-smash, the cyclorama sky bordered by cat tails and reeds the Duck used as the grand curtain when he played the Palace. He SLAYED 'em, always, though a little bill-bashing was necessary. Honked Fudd's taxi-bulb nose and swan dived (saying so) onto the water now like obsidian KRR-AKKK! Orchestral cymbals for his lying down. Recovered to read a sign near the shore, HARD WATER. *Woo-hoo!*

From the pond a beaver stream became the river watering frontier pictures. Beavers picked up and used as power tools, though the Duck and the Rabbit took little interest in that kind of thing, they're not MEAN TO OTHER ANIMALS like that damned dirty Mouse. Fudd as camper. One thousand tree gags. Tree with an elevator in it. The beaver stream quickened: a waterfall. Close calls on jostling cakes of ice, logs, perils of the silents rendered in undertones of Currier & Ives. Every stop of bonnet-and-sawmill melodrama. The Rabbit held out a branch at the brink of the hereafter, snatching it away from Fudd or the moustachio'd redhead—*Ain't I a stinker?* Waterfalls could be run up, or even rowed back up, vertically, this blue water.

From the waterfall and the chaos up at the top there, a river now—powerful enough to generate voltages for all the Tesla coils in the evil scientist's cracked baronial (the Rabbit had a few scrapes there), power assembly lines in numerous factories (odd how the Duck and others often had to fall back on factory work—the War again), power the cannery where almost everyone got minced or bottled one time or another. Power for the heroine-sawing mills, the House of the Future gone wrong, the lights of vaudeville and Hollywood. Long rectangular-paned windows of the white modern world: powerhouse, factory, Dad's Lab. Industry was never threatening here. Oh you might get snatched up and temporarily jerked around in some mechanical-arm, conveyor-belt way. White of the lab coats the same as the walls, sunlight streaming in, many of our toys hailed from here—Gilbert, Ideal, Remco. Bubbling retorts, condensers, battalions of test tubes, evil mixtures which gave the Duck crackled bloodshot eyes and green feathers, turned the Rabbit plaid. Their glowing fumes cast garish shadows. Maybe this is the Science I love, not Dad's, not Herbert S. Zim's.

They were always visiting the Gay Nineties. Ball-headed men with petrolatum moustaches, ladies in bustles, sleighs decorated everything, Jack Hass's bar by his pool on the other side of the fence was like this. All the printed trash of Christmas. Old New York's appeal to bumpkins in barnyards, cowlicks and slickum, wet-behind-the-ears kids and the high collar Rabbit going there for the first time, *Look—I'm dancin'!*

Always the roots of show business. Poolside the Rabbit recounted his career, maybe burlesque, vaudeville. Benny, Cantor, Bing, they were all there, suffering lean times in frozen Central Park where the Singing Frog and his impresario spent the winter. *Bugs Bunny! What are you doing hanging around with these bums? They'll never amount to anything!*

The Rabbit in spats. Neighborhood bulldogs in turtleneck sweaters, grey derbies, stogies. East Side, West Side, the 'Daughter of Rosie O'Grady', barber shops with hot towels, with hot-towel-things-that-look-like-English-armor, as Julie would say. The haunts of men together. Organ grinders. Where else but from this vaudeville would Julie and I know that to act NONCHALANT you stuck your hands in your pockets and whistled 'While Strolling Through the Park One Day', looking from cloud to cloud? Just as I learned Stan Laurel's look of pathologic contrition which made Dad want to hit me on the head, he

179

said. How else would we know you had to hum the *Erlkönig* when you're in danger? One summer evening as fast as I could make holes for eyes in a series of brown paper bags, put them over my head and run out to the street, they were ripped from my head and shredded by Larry, he was hysterical, I'm from another planet, I said, *RIP!*, I'm from another planet, *RIP!*, from the Rabbit and silent comedies at five a.m. on Saturday I found the mad repetition of this. *You're overstimulated*, said Mom. Julie and I used up a whole can of shaving cream and a pack of paper plates, I mounding up phony pies, she mashing them into my face, why you, BAM, why you, BAM. Where else would we have seen a sailing ship, a Pacific locomotive, heard of Pennsylvania at all, or Sing Sing?

There was the new city, the midwest via Los Angeles, thrusting blocky buildings. Through their roofs the Duck and Rabbit often crashed in ongoing elevators, or tricked each other off them. The Duck in cape and long johns jumped over these martini buildings as did the Rabbit after eating SUPER CARROTS from the white laboratories. Traffic music, Gershwin, Grofé, busy violins, the four-note horns of long limousines turning corners in the liquid middle of themselves, moop MOOP moop MOOP!, the trolley gong, stoplights with bells, stoplights which reached out white gloved hands to transgressors, *Hey, Mac! Where do ya think YOU'RE going?*, stoplight cousins of the factory whistles that put two white gloved fingers in their mouths and blew.

You know how to whistle, don't you, Steve? The city was bright and angular, shed of its Buicky big band bulges, new TREES stylized in penwork freely splashed with watercolor greens. The stout barn-red hydrants. This was the city Julie and I were handed in our books, clean white schools with large windows and stair-towers, schools like the SUPER CARROT laboratories, schools like in *Arithmetic Town*.

You could watch the nation from above, watch a bold red line chart a crazy journey, see everyone running all over the map in search of something, see the Duck smash into Mount Rushmore or a dog come all the way from Juneau stop, panting, beside a sequoia. The limitless continent. Julie and I had been out on it, we knew the colors of cañons, of beaches, of redwoods. Not that we could always see them, not that we were tall enough to see once our kiddie seats were gone. But we knew these colors, knew the continent from postcards we collected, brochures brochures brochures, huge racks of them overflowing gas station and Winston-smoked motel lobby. Collect-em-all, our ethic.

Giant redwoods, their trunks rich brown against their foliage and gold sand by the Eel River, the relieving posts for packs of nutty dogs. *There's that pooch again—I wonder what he's after?* Mr Avery must have grown up across the street from the most dog-popular hydrant in Texas. Beaches licked by the slinky denim blue, umbrellas slanted this way and that into the distance, legs, the ends of beach chairs, dark green primary-striped canvas, sand pails. The everywhere-beach of colors begun in the pamphlets of 1900. A cross-eyed shark sneaked up behind the Cat, opening a mouth as wide as Sunset Boulevard with everyone swallowed at Warners since 1929 beckoning in his innards. Sometimes his teeth were briefly xylophonic.

They made one giant National Park with features of every one. The Rabbit and Fudd chased through Yosemite's greens, played infinite tricks on each other at Old Unfaithful's blow-hole, Old Faithless which rumblingly mightily shook the earth only to PUT! DING! one lousy little gob in a wee spittoon. Or else blasted Fudd sky high and kept him there. All of his picnics gone wrong, ants who had a President and wore GI helmets attacking the food, sandwiches and jars scuttling away on little legs. Boxes of barn-red Fourth of July fireworks Jolsonized Fudd's face. Tortuous roads up Pike's Peak, a Mount Rushmore that talked, bear caves with Murphy beds.

The desert sanded the edges of the plains. Fudd, the Rabbit, the Cat going west how many hundreds of times, leaving giant everywhere-Ohio, leaving the daughters of Rosie O'Grady, past the improbable Yorktowns, past the hollers where Martins and Coys did things to each other, *Arkansas Traveler*, rustic bridge, corn likker jug XXX. The desert endlessly repeating. How many times did the Rabbit don the bleached skulls of oxen and scare the pants off someone, how many cattle died out there anyway, at the foot of every cactus a cinnabar boulder and skull? And how many humanoid saguaros flung their needles in anger? The mirages, Julie and I strained in our kiddie seats to see them. I thought once we reached the desert they would be everywhere. I didn't know they were puddles of shimmer, I thought they would have towns and dancing girls. By the fifties the Rabbit's mirages were packed with hamburger stands, swimming pools, Palm Springs hotels. The desert was how you got to the Old West, dog families in wagons heading away from the deltas of Mighty Mouse's steamboat melodramas. Every Conestoga and frying pan gag.

Plink plinkitta plink plinkitta plink plink plink plink plink plink, coming into the frontier town past signboard gags, stoplights with bells which halted fusillades of bullets criss-crossing the only junction. RIGOR MORTIS SALOON—COME IN AND GET STIFF. The undertaking dog with top hat and tape measure. And boy did they know how to build a fort, of browns browner than the browns you dozed over in school books, the redwood forests, even the peculiar watercolor box of Herbert S. Zim, the logs of the stockade sharpened like 3B pencils. The Duck in a huge bicorne, the sign of utter madness. Suddenly, tom-tom and oboe! Big fat Indian in a war bonnet, scanning the horizon, as he turned you saw it was a turkey on his head, also scanning the horizon, wing to brow. Indians crawl oozy like snakes. Indian telephone no bell, tom-tom instead. BOM bom bom bom. Brave on bicycle, not horse, grindum tomahawk on front wheel. Navajo Morse from a giant Havana. Thousands of braves erupted on horseback from one teepee. They had arrow revolvers, loosed an entire quiversworth at once, were shot and spun around under their horses, righting themselves uninjured, the politics of a shooting gallery. The Duck swallowed black powder and balls. The Pig seized him and used him as a blunderbuss.

Big Chief Rain-in-Face, under a constant downpour from his personal thunderhead which followed him around, promised the hand of his gorgeous daughter to the brave who defeated the Duck and the Pig. No brave ever see Princess Laughingwater face behindum turquoise veil but WATTA BILT.

The girls. The gorgeous skin-sloughing reptiles and amphibiennes in the phylum nightclubs of the animal kingdom, the Penguin Club, the Oyster Bar. Betty Bacall was their favorite, according to Dad when he ever watched. Once in a while an evil genius constructed in her image a bumping grinding ROBOT FEMALE with a bomb ticking inside her, whose Max Factor thumbprint lips delivered 50,000 VOLTS, what a woman, to the smitten Rabbit or Duck, who developed a dangerous-looking GIANT CONGA HEARTBEAT, the red outline of the thing railing against their ribs.

The housewives, scolds of apron and of leg, seen from the floor. We all live on the floor. They polished and scrubbed their kitchens of light and chrome and sea-green linoleum, like the white laboratories, and in their white refrigerators was the plenty of the farmer on his tractor in the cornucopia of the dark: armories of golden carrots with lush green shaws like the ones Mom chopped for us in her kitchen now, for watching the Rabbit.

Flying above the limitless continent with Julie, I looked down on the big colorful nation: the brown old towns, the white modern cities, the plains, the rivers, the TREES. And Fudd, Hitler, the Kellys, Nunzio, his mom, the Pupae, Gomez, Larry, Mrs Plank, Principal Dummkopf, ARITH-METIC—outwitted, ridiculed, foiled, insulted, hit on the head with a gigantic mallet; killed. Made very flat with a steamroller or chased, zig-zag, off over the horizon.

Never be ashamed of kindergarten—
It is the alphabet's only temple.

BILLY COLLINS, Instructions to the Artist